BESTSELLING BOOK SERIES

RFID For Dummies®

W9-BSQ-975

Cheat Sheet

Helpful RFID Web Sites

General RFID Information

Auto-ID Labs	www.autoidlabs.org
Auto-ID Lab @ Adelaide	http://autoidlab.eleceng.adelaide.edu.au
IDTechEx	www.idtechex.com
RFID Exchange	www.rfidexchange.com
RFID Gazette	www.rfidgazette.org
RFID Journal Online	www.rfidjournal.com
RFID Solutions Online	www.rfidsolutionsonline.com
RFID Update	www.rfidupdate.com
Slashdot	http://slashdot.org
UCLA's RFID @ WINMEC site	www.wireless.ucla.edu/rfid/research

RFID Standards and Protocols

Automotive Industry Action Group (AIAG)	www.aiag.org
EAN.UCC	www.ean-ucc.org
EPCglobal	www.epcglobalinc.org
International Electrotechnical Commission (IEC)	www.iec.ch
International Organization for Standardization (ISO)	www.iso.org
UCCnet	www.uccnet.org

Project Lifecycle Checklist

- ❑ 1. Identifiy critical stakeholders and form an RFID team.
- ❑ 2. Refine the process and conduct team training.
- ❑ 3. Determine scope and assumptions.
- ❑ 4. Determine drivers, strategies, and enablers.
- ❑ 5. Identify and assess business processes and interfaces.
- ❑ 6. Identify complementary or competing business initiatives.
- ❑ 7. Identify strategic and economic benefits.
- ❑ 8. Develop investment requirements.
- ❑ 9. Develop an implementation road map.
- ❑ 10. Communicate the business case.

For Dummies: Bestselling Book Series for Beginners

BESTSELLING
BOOK SERIES

RFID For Dummies®

Cheat Sheet

RFID Interface Protocols

Protocol	Corresponding Frequency	Capabilities	Pros	Cons
Generation 1 Class 0	UHF	This is a preprogrammed tag, which means that the end user can't write a new number to the tag.	Slightly lower cost. Good overall performance.	Preprogrammed tags can increase administrative and logistics cost. Proprietary standard.
Generation 1 Class 1	UHF and HF	Write once, read many (WORM)	Keep data in sequential order; manage data easier. Open standard.	Can be written to only once.
ISO standard 18000	LF, HF, and UHF	WORM	Keep data in sequential order; manage data easier.	Does not account for the data structure but only how the tag and reader communicate.
Generation 2.0 Class 1	HF and UHF	WORM	Keep data in sequential order; manage data easier. More globally accepted protocol.	Can be written to only once.

Frequencies, Power, and Countries

The following table lists the frequency bands used in different regions around the world.

	UHF	HF	LF
North America	902–928 MHz (4 watts of power, 4W ERP)	13.56 MHz	125–134 kHz
Australia	918–926 MHz (1 watt of power, 1W ERP)		
Europe	865.6–867.6 MHz (2 watts of power, 2W ERP) 865.0–865.6 MHz (0.1 watt of power) 867.6–868.0 MHz (0.5 watt of power)	13.56 MHz	125–134 kHz
Japan	950–956 MHz	13.56 MHz	125–134 kHz
	(**Note:** The allocation is temporary as Japan tries to get closer to U.S. and European standards.)		

For Dummies: Bestselling Book Series for Beginners

RFID FOR DUMMIES®

by Patrick J. Sweeney II

WILEY

Wiley Publishing, Inc.

RFID For Dummies®

Published by
Wiley Publishing, Inc.
111 River Street
Hoboken, NJ 07030-5774
www.wiley.com

WILEY

About the Author

As you may have guessed by the dangling participles and misused gerunds, this is the first book by **Patrick J. Sweeney II** (despite Amazon's link to books on gynecology by an author of the same name). When not negotiating with his editor to push back book deadlines, he leads ODIN technologies as President and CEO.

ODIN technologies is a global RFID software and services company focusing on RFID infrastructure. Mr. Sweeney is well recognized as a visionary in the RFID industry with several RFID patents in various stages of approval. He has appeared in such publications as *CIO Magazine, The Washington Post, Fortune* magazine, *Internet Week,* and many others. He has been interviewed by ABC news and CNN, among others, and is a frequent speaker worldwide on all topics relating to RFID. He is also an active member of several standards bodies and regulatory groups helping to shape the evolution of the RFID industry.

Mr. Sweeney is a second-generation IT professional; his father was one of the first employees at Electronic Data Systems (EDS), where "Pops" entertained him and his brother on weekends by teaching them to read punch cards and other useful skills. Mr. Sweeney took that genetic proclivity toward data centers and started a successful, secure managed hosting company in the late 1990s, which he later sold. His brother took that same early training and started XS Speed Choppers, making custom motorcycles — go figure.

Mr. Sweeney finished second in the 1996 Olympic trials in the single scull, is an avid outdoorsman, enjoys helping other entrepreneurs, and is passionate about various Irish causes. He is a board member of Trinity College business school in Dublin, Ireland, and an Alumni Board member at the Darden School of Business at the University of Virginia. He graduated from Darden and received a Bachelor of Science degree from the University of New Hampshire. He is blessed with a great family—wife Christen, daughter Shannon, son P.J., and three dogs. They live in Middleburg, Virginia, in a house full of useless RFID gadgets.

Dedication

This book is dedicated to everyone who makes the dream of entrepreneurship and innovation possible, from the brave men and women defending our freedom in the armed forces, police, and fire departments to college professors, mentors, and angel investors.

Topping the list of people who make entrepreneurship (and crazy book projects) possible are loving, understanding, and helpful spouses like mine. This book is especially dedicated to my beautiful wife Christen, who helps and supports me as I build companies, write books, and travel around the world chasing birds and the Red Sox.

Author's Acknowledgments

First and foremost my family gets a big thanks for letting me bang away on the laptop during dinner, in bed, and at other times made awkward because an electronic device is the center of my world instead of them. Thanks Betty, Shannon, and P.J.!

The book flow, formatting, and funniness (I recently learned that there is called *alliteration*) is largely due to the great work of Becky Huehls at Wiley who was my project editor and learned me all sorts of interesting things about writing.

Of course the book wouldn't even be possible if not for the guys in ODIN technologies labs; Bret, Charles, Nick, Ray, Dave, and the rest of the crew played an invaluable role, and they deserve a ton of updog.

I could not have written such a comprehensive book on this diverse technology without significant contributions from some first-class industry experts. Many of these folks contributed an entire chapter to the book, so although the pronoun "I" is used throughout the book to stay consistent with Wiley's *For Dummies* style, much of the credit goes to an amazing team of contributors:

Earl Cox
Scianta Intelligence
www.scianta.com
www.autoidlabsus.org

Daniel Engels, Ph.D.
Robert Goodman
Yankee Group
www.yankee.com

Pat King, Ph.D.
Bob Brescia
Michelin US
www.michelin.com

Sharyn Leaver
Forrester
www.forrester.com

Chris Fennig
ODIN technologies
www.odintechnologies.com

I also thank God for blessing me with great family and friends who helped me get to a position where taking on this project became a reality. Thanks Mom and Pops, Blanche, MAF, Jimbo, Shelley, Rusty, Vas and Linda, Chris and Kate, Gregg, John M, David B, Robert, Zohar and Sam, Bernard, Charles, Murph, Melchoir, Bo, Dr. R, and everyone else who helped me get here.

Publisher's Acknowledgments

We're proud of this book; please send us your comments through our online registration form located at www.dummies.com/register/.

Some of the people who helped bring this book to market include the following:

Acquisitions, Editorial, and Media Development

Project Editor: Rebecca Huehls

Acquisitions Editor: Melody Layne

Copy Editor: Andy Hollandbeck

Technical Editor: Christopher Bratten

Editorial Manager: Leah P. Cameron

Media Development Manager: Laura VanWinkle

Media Development Supervisor: Richard Graves

Editorial Assistant: Amanda Foxworth

Cartoons: Rich Tennant, www.the5thwave.com

Composition Services

Project Coordinators: Adrienne Martinez, Emily Wichlinski

Layout and Graphics: Carl Byers, Andrea Dahl, Lauren Goddard, Denny Hager, Joyce Haughey, Lynsey Osborn, Melanee Prendergast

Proofreaders: Laura Albert, Leeann Harney, Jessica Kramer, Linda Morris, Carl William Pierce

Indexer: TECHBOOKS Production Services

Special Help: Kim Darosett, Teresa Artman

Publishing and Editorial for Technology Dummies

Richard Swadley, Vice President and Executive Group Publisher

Andy Cummings, Vice President and Publisher

Mary Bednarek, Executive Acquisitions Director

Mary C. Corder, Editorial Director

Publishing for Consumer Dummies

Diane Graves Steele, Vice President and Publisher

Joyce Pepple, Acquisitions Director

Composition Services

Gerry Fahey, Vice President of Production Services

Debbie Stailey, Director of Composition Services

Contents at a Glance

Table of Contents

Introduction

Somewhere, separated from you by just a few degrees, is not Kevin Bacon, but an 800-pound gorilla demanding that you adopt radio frequency identification, or RFID — a technology you may have never even heard of until just a few months ago. Chances are that gorilla wears a stylish blue smock with a yellow smiley face on it and greets you with a "Welcome to Wal-Mart." If not Wal-Mart, the US Department of Defense, Target, Albertsons, Best Buy, Tesco, Metro, the FDA or a number of other companies may be requiring you to implement this technology by a certain deadline. If you don't have a mandated deadline for adopting RFID, consider yourself lucky. You can discover and make decisions about this exciting technology based on your normal process for evaluating new business tools.

Whatever your situation is, you either want or need to set up an RFID network. So you went out and picked up *RFID For Dummies* and are ready to go — yippee!

About This Book

This is a book that is on a mission to take the confusion out of RFID. RFID is based on well-known laws of physics. It's easy to understand how things work after you get your arms around those basics. The better news is that the technology works really well if you know what you're doing. So without sending you to MIT for a couple of years of RF engineering school, this book explains everything you need to know to start setting up and deploying your own RFID network — what more could you ask for?

Who This Book Is For

Whether you are just curious, scared, worried, or simply mad at the prospect of implementing yet another new technology — even if you know nothing about RFID — *RFID For Dummies* is here to help. And, unlike a similar promise by the IRS, this book really will help. You find out what RFID is, what it does, and how it works. I guide you through the concepts and ideas in plain English, walk you through the basics of RFID from a business perspective, and speculate on where this technology is headed (although I do, from time to time, provide sufficient Geek Speak for the engineers and systems guys who, no doubt reluctantly, bought this book in an attempt to actually understand the mechanics of Radio Frequency Identification).

If you know the basics about running a laptop or PC and know what an IP address is, you are armed with just about all you need to know to initially set up an RFID network. If you have any background in physics and understand some things from an electronics perspective, you've got a running start. I assume that you come from a supply chain or warehouse background and might not have a detailed IT background.

You Don't Need a Slide Rule and Pocket Protector to Use This Book

Other than the willingness to learn and basic knowledge, you need some equipment to set up your RFID network and follow some of the processes outlined in this book. At some point, plan to get

- ✔ A spectrum analyzer (discussed in Chapter 8)
- ✔ A budget to buy an RFID reader, antennas, tags, and a rack (about $7,500 total)
- ✔ An area large enough to begin testing and using the equipment (at least 20 feet x 20 feet)
- ✔ Another person to help you occasionally try out the technology
- ✔ A penchant for experimentation and thirst for knowledge

How This Book Is Organized

RFID For Dummies is broken into six different parts. If you are new to the technology, it is helpful to read the parts in sequential order. If you have a physics or RF background and you want to get into the nuts and bolts of the technology, skip right to Part II and then move on to Part III. If you are trying to justify the RFID project, you may want to go right to Part V, which addresses some of the business concerns around strategic planning and ROI. You can read all the technical chapters in Parts II and III by themselves and use them for reference, as well as the last part, the Part of Tens. Here's a quick rundown of what you'll find in each part.

Part I: Now That You Can Spell RFID, Here's the Rest of the Story

This part introduces the basics of RFID. In Chapter 1, you find an overview of the technology, what advantages are driving the mandates, and a blueprint for implementing RFID, which I call the four Ps. In Chapter 2, I explain how RFID fits into the world of Auto-ID technology and explain some of the basics about the protocols that make it work. Chapter 3 helps you start assessing the impact RFID will have on your business and helps you make some basic decisions about how you'll use RFID.

Part II: Ride the Electromagnetic Wave: The Physics of RFID

In this part, I peel away the layers of RFID to uncover the underlying science of RFID. This part gives you the physics knowledge you need in order to design your network for optimal performance and make wise purchases. In Chapter 4, you can find an overview of how the physics of RFID systems work. Chapter 5 digs a little deeper by delving it parts inside each of the key components of a system. Whereas Chapters 4 and 5 focus on the invisible realm of electromagnetic waves, Chapter 6 is focused squarely in warehouse or marketplace, covering common setups of RFID systems and case studies so that you can learn from early adopters.

Part III: Fitting an RFID Application into Your World

This part is your key to designing an RFID network specifically for your environment and needs. In Chapter 7, I walk you through the process of testing for electromagnetic noise in your warehouse or building using a spectrum analyzer. Chapter 8 helps you set up a lab (or find one you can use) so that you test for the right tag (Chapter 9) and tag reader (Chapter 10). And last but not least, Chapter 11 helps you wend your way through maze of *middleware* (the software the connects the RFID network) by explaining what features to look for and how to fit middleware into your network architecture.

Part IV: Raising the Beams for Your Network

This part walks you through the process of actually implementing your carefully planned-out RFID network. Chapter 12 explains a few project management tools that will keep your trial run and follow-up network designs on schedule. Chapter 13 covers the process of setting up the hardware in the warehouse, or other real-world setting (as opposed to a lab), and how to train your employees to use the new system. And Chapter 14 explains now to set up monitoring systems for both operators and system administrators, so that your system keeps running strong, and thus helps your bottom line.

Part V: How to Speak Bean Counter

Deploying an RFID system is a big project, and the bottom line needs to drive your implementation. This part walks you through the key RFID-related business decisions you need to make. In Chapter 15, I explain who in your organization needs to be involved in these decisions and walk you through a nine-step process for building and presenting a business case study. In Chapter 16, I explain strategic benefits you need to include in the business case in more detail, including how to calculate return on investment, or ROI, for all the money you're about to spend on RFID hardware and software. Chapter 17 is your guide to outsourcing: I explain how you decide whether to outsource, what to look for in an outsourcing partner, and how to seal the deal.

Part VI: The Part of Tens

No *For Dummies* book is complete without a Part of Tens. The four chapters in this part offer (more or less) ten equipment vendors to assess, ten of the best RFID-related Web sites, ten tips from RFID experts who are part of that rare fraternity that has actually done real-world deployments and lived to tell about it, and ten standards and protocols for RFID that you may want to investigate.

In the back of this book, you can also find a glossary of electrical, magnetic, and scientific terms. So if, in your RFID reading, you come across terminology that leaves you baffled, you can use this glossary as a handy resource.

Icons Used in This Book

Throughout this book, you find icons in the margins, marking specific paragraphs. Here's what those icons indicate:

The Tip icon marks tips and shortcuts that you can use to make your RFID installation, testing, and implementation easier.

Remember icons mark the information that's especially important to know. To siphon off the most important information in each chapter, just skim through these icons.

The Technical Stuff icon marks information of a highly technical nature that you can normally skip over unless you have a closet desire to geek out on radio frequency. But face it: If you're reading about RFID, you're probably a technical-minded person. If this is the case, you're more likely to skip *to* this icon instead of skipping *over* it.

The Warning icon tells you to watch out! It marks important information that may save you headaches, long talks with government officials, and maybe even bodily injury.

The Case Study icon points out real-life examples of how RFID has been used (and misused) in the marketplace.

Part I

Now That You Can Spell RFID, Here's the Rest of the Story

The 5th Wave   By Rich Tennant

TARZAN – LORD OF RFID

"... and then one day it hit Tarzan, Lord of Jungle – where future in that?"

In this part . . .

Part I gets you prowling down the path to RFID adoption. In these three chapters, you become acquainted with the basics of the technology and understand how it compares to other automatic identification (Auto-ID) technologies. I explain why RFID has blossomed into the latest and greatest technology since the electric toaster. You also find out why so many people need to adopt this technology in such a short period of time.

The last chapter of Part I shows you, in simple, easy-to-understand terms, how to compare the different RFID networking and technology systems. This serves as a primer for more detailed discussions later in the book.

Chapter 1

Taking the Mystery out of RFID

*W*ith all the recent hype over radio frequency identification (RFID) and the requirements to implement it, you might think that RFID can turn water into wine, transform lead into gold, and cure the world's diseases. You might also be worried that RFID will enable Big Brother to track your movements to within a foot of your location from a satellite five hundred miles up in space. The truth is, RFID can do none of these things.

In this chapter, you find out the basics of what RFID is, what forces are driving RFID as a replacement for the bar code in the marketplace, and what benefits RFID can offer.

If you are responsible for complying with high-profile mandates from one of your suppliers or customers, this chapter also offers a framework to help you begin setting up a system and making it work within your existing business process. The bad news is that an RFID implementation is a daunting project even at a minimal compliance level, sometimes referred to as *slap and ship* or, more appropriately, *tag and ship*. The good news is that the benefits to the business are substantial, particularly if your trading partners are involved. RFID technology is here to stay, so the sooner you understand it, the quicker you can make key strategic decisions for your company.

What Is RFID?

RFID is a very valuable business and technology tool. It holds the promise of replacing existing identification technologies like the bar code. RFID offers strategic advantages for businesses because it can track inventory in the supply chain more efficiently, provide real-time in-transit visibility (ITV), and monitor general enterprise assets. The more RFID is in the news, the more creative people are about its potential applications. For example, I recently heard from someone who wanted to use RFID to track fishing nets in the North Sea.

The origins of RFID in inventory tracking

Wal-Mart has spent millions of dollars since the late 1990s researching the efficacy of RFID systems to replace bar codes (which have been in use since the days of *The Brady Bunch* and *Gilligan's Island* — that's the early 1970s, for those of you with all your hair left).

In 1999, with the help of scientists at the Massachusetts Institute of Technology (MIT), a consortium of companies formed the Auto-ID Center — a center for continued research into the nature and use of radio frequency identification. The consortium had a new idea about how organizations could identify and track their assets. The vision underlying automatic identification (or Auto-ID) is the creation of an "Internet of Objects." In such a highly connected network, devices dispersed through an enterprise can talk to each other — providing real-time information about the location, contents, destination, and ambient conditions of assets. This communication allows much-sought-after machine-to-machine communication and decision-making, rendering humans unnecessary and mistakes a thing of the past.

Today, Auto-ID can track not only enterprise assets, but also the movement of products, containers, vehicles, and other assets across vast geographic areas. For more about the Auto-ID Center and the current organizations involved in developing RFID technology, see Chapter 2.

Tracking goods with EPC codes

RFID is actually nothing new. Just as goods today have bar codes, goods in RFID systems have codes that enable systems to share information. Because the mandated RFID systems require businesses to share information with each other, the different systems need to use the same code — the electronic product code (EPC). The EPC is the individual number associated with an RFID tag or chip.

The EPC was developed at MIT's Auto-ID Center in 2000 and is a modern-day replacement for the Universal Product Code (UPC). A tag's embedded EPC number is unique to that tag. However, the EPC *protocol* is universal to all EPC-compliant systems and serves two specific functions:

- Telling how data is to be segregated and stored on the tag, or what is also known as the *numbering scheme*.
- Determining how the tags and readers communicate (also called the *air interface protocol*).

Wal-Mart, like other large retailers, had more pragmatic issues at hand when they established an RFID requirement for their suppliers. Under Wal-Mart's mandate, each supplier is required to identify their products not by bar codes and waybills, but through EPCs that are automatically broadcast by RFID tags as new products arrive at the retailer's warehouse, distribution center, or store. In Chapter 2, I explain how EPC works in more detail.

Sizing Up the Benefits of RFID

Capturing inventory as it arrives from the supplier is the first step in a company-wide tracking system that "knows" where every item is throughout its lifetime in the store. This tracking offers retailers tremendous insight into their inventory, which enables those retailers to control costs and reduce investment on inventory, which means lower prices and better competition for consumers.

Having better information about inventory offers retailers all sorts of potential benefits. The retailers know how much inventory is still on pallets in the warehouse, how much is on its way to distribution centers and stores, and how much is currently on the shelves in each of its stores. With this knowledge, retailers have the foundation for measuring product consumption, seeing buying patterns, and controlling inventory more efficiently. Through this process, a retailer ensures that its shelves are stocked and that customers can buy high-volume products (such as razor blades, diapers, and toilet paper) when they need them and in the quantity they need.

Of course, businesses don't spend money unless they expect to make money off that investment. Major retailers believe that a comprehensive RFID program — tying suppliers to inventories to retail outlet shelf stock — will generate savings of around 10 to 16 percent, based simply on inventory cost reduction in each of their distribution centers (DCs). This translates into billions of dollars in savings each year — a pretty impressive result by any measure. The benefits can extend to other applications beyond retailers: Third-party logistics companies can speed up their billing cycle and create a new revenue stream with RFID; government agencies can reduce loss and increase security; museums can reduce cost to conduct inventory; sports teams can increase sales at games — the applications are limitless.

In an RFID system that uses an electronic product code (EPC) or similar numbering scheme, the following RFID attributes lead to those kinds of savings:

- **Serialized data:** Every object in the supply chain has a unique identifying number.

- **Reduced human intervention:** RFID allows tracking automatically without needing people to count or capture data or scan bar codes, which means reduced labor costs and fewer errors.

- **Higher throughput supply chains:** RFID allows many items to be counted simultaneously.

- **Real-time information flow:** As soon as an item changes state (off the shelf, out of a truck, sold to customer), the information can be updated across the supply chain.

- **Increased item security:** Tagging items allows them to be tracked inside a confined facility or space.

In the following sections, I explain each of these benefits in more detail. In Chapter 2, I compare RFID to other auto-identification technologies, like the bar code, and offer tips for developing an overall Auto-ID strategy so that you see how you might apply RFID's benefits to your own business.

Obviously, there is a genuine reason for the excitement surrounding RFID and the EPC. People are anxious to implement the technology so they can track supplies from the factory to the foxhole, or from the grower to the grocer. Much like the excitement surrounding the Internet, RFID carries the promise of a very disruptive technology with substantial future rewards. The excitement (dare I say *hype?*) needs to be tempered by the real-world limitations of the technology and the laws of physics. Adding to the practical limitations of today's RFID technology is a deluge of misinformation and broken promises. Today's marketplace dynamic is the cause of much of this RFID heartache. I introduce a well-balanced approach to RFID in "Finding Success with Four Ps in a Pod," later in this chapter, to make sure that you stay on an even keel and take a pragmatic, process-driven approach to the technology.

Tracking individual items with serialized data

Serialized data means that each item has its own unique identifier or serial number. This helps an enterprise

- **Keep very accurate account of each item in the supply chain or property list.** Instead of knowing that there are 1,000 boxes of Cap'n Crunch (get it? *serialized* data) in the back room, a grocer knows which box has been sold and which one has been sitting there for a long time.

- **Know which item was produced where, in companies that produce the same item at multiple plants.** This is critical for tracking total quality, aiding in recalls, verifying warranties, and so on.

- **Prevent counterfeiting and diversion.** Serialized data allows items such as high-cost drugs to travel through a supply chain while recording every stop they make.

The benefit of serialized data is better inventory control, reduced loss, reduced carrying cost, and improved customer satisfaction (customers at every level, not just walk-in-off-the-street Joe Brown). Each of these advantages over the existing system has a benefit of reducing cost and improving productivity (another way of saying the same thing!).

RFID tracks individual items by associating the unique EPC number to a secure database. This concept is often likened to license plates. Just like the DMV knows who owns a car by looking up the license plate number on a central server, an RFID system can pull up a limitless amount of information about a tag based on its unique identifier.

In some instances, particularly with active tags, the RFID tag allows all the critical information to be stored directly to the tag. No need to look to a database — all the info is right on the tag. This technology can be very useful in instances such as the shipment of military supplies to overseas theaters, where accessing a central database is nearly impossible.

Reducing human intervention

Thousands of applications require humans to scan an object with a bar code scanner or read information on a label. When you check out at the supermarket, the checker has to pass each item in your cart over the lasers that scan the bar codes. RFID technology has the potential to eliminate this human intervention. If all your groceries had RFID tags, you could walk straight out the door and have all the items in your basket read automatically as you passed by a portal, with no need to take things out and scan them.

Think about cases of items coming off of a tractor trailer into a distribution center. Today, someone scans each box one at a time with a bar code scanner and often sticks a label on the box as it leaves the truck. From a logistics perspective, RFID can automatically verify a shipment, optimize cross-docking and flow of goods, and automate much of the pick-and-stow functions. With RFID, things can move off the truck by the pallet-load. Hundreds of items can be read simultaneously, and the data can immediately hit the inventory system as being on-site, identifying what it is, where it came from, where it's going, and so on.

The benefit of having fewer human hands involved is reduced errors, which produces reduced costs, faster throughput, and reduced damage and returns. The overall implication of reduced human intervention, given the high cost of salaries, benefits, and the cost of management associated with crews of human workers, is a dramatic reduction in operating costs.

Automated toll systems are a prime example of how the lack of human intervention saves both time and money. Remember how long the lines at highway tollbooths used to be? This was especially annoying if your daily commute was on a toll road. With automated toll systems (made possible by RFID), no longer does a car have to stop to hand cash to an exhaust-inhaling person stuffed in a 2-x-3-foot box all day. Zoom by and smile. Less traffic, lower cost, elimination of a hazardous job. Thank you RFID!

Moving more goods through the supply chain

Supply chains that can move more goods (also called *higher throughput supply chains*) reduce processing time, which leads to reduced costs, higher turn-around for billing customers, improved cash flow, a better bottom line, and, of course, reduced error rates, which also contribute to improved customer service. This leads to better customer retention, higher sales, and an increase in profitability and throughput performance.

Before RFID systems became a viable Auto-ID technology, systems with high-volume throughput (airline luggage handling, package delivery, road race participants) all had to be read one item at a time because a bar code scanner can read only one bar code at a time. Whenever only one item is read at a time (manually or with a bar code), the maximum throughput is — you guessed it — one.

Entire systems were designed around processing *one* as quickly as possible. Fred Smith, the CEO of FedEx, spent millions trying to figure out how to collect one package at a time and read it in the shortest amount of time as it goes down a very high-speed conveyor. That was the design goal of systems that required optimization of a one-at-a-time bottleneck.

RFID changes all that by allowing a whole bundle of packages, a trailer of luggage, or tens of runners to be read all at once, greatly increasing through-put. With RFID, you can read hundreds of objects all nearly simultaneously. No longer will systems be designed to optimize the speed of *one;* rather, they will be designed using the laws of physics to maximize the number of simultaneous reads.

Capturing information in real time

Real-time information can help you reduce costs, improve sales, increase cash flow, allow for specialized servicing and manufacturing for top customers, and thus capture a larger market share and improve overall capitalization per client and per employee. Because you know, in real time, where everything is, you can deliver on promises, reduce errors, increase customer loyalty, reduce waste, optimize materials use, and directly impact the tactical (departmental) and strategic (corporate and division-level) bottom line.

If time is money, information is insurance. What is on a store shelf, off the shelves, selling well, about to spoil, running low in back, and missing is all critical information to a retailer, producer, or supplier.

An RFID system can also allow machine-to-machine communication and automated decision-making. Automated decision-making is based on two principles of RFID: lack of human intervention and real-time information flows. In real time, a conveyor can close a gate and route a package at 600 feet per minute from one line to another line all because it reads the data off an RFID tag and retrieves a command specific to that individual item (it's that serialized data benefit again).

Increasing security

RFID's increased security means improved delivery and control and increased anti-counterfeit measures, as well as theft reduction, which leads to a significant reduction in costs.

If you are responsible for the tracking and accounting of property items, or if shrinkage to you is more than what happens when you jump into that frigid Cape Cod Bay, RFID is a dream come true. (*Shrinkage* in an inventory sense is the loss or theft of items in the supply chain.) The ability to permanently affix a tag to every item of value in a location and know exactly where that item is at all times as it passes through various doorways is something no other technology can offer. From a security perspective, RFID's ability to track and trace property can help everything from the war on terrorism to anti-fraud and anti-counterfeit measures. Here are some examples:

- The pharmaceutical industry not only deals with fake drugs being passed off as the real deal, but is fighting a multibillion-dollar issue of *diversion*. Drugs have different price scales for different buyers. Distributors know who pays less for drugs — like hospitals and nursing homes — and some less-than-upstanding distributors take advantage of these price differences to illegally turn a profit. See Chapter 6 for more details.

- *Gray market* items (items that are made in the same plants or with the same markings as a real product but sold much cheaper on the black market) are another problem easily solved with RFID: Embed a chip in every Fendi bag and you'll be able to tell the fake ones sold on the street from the real ones sold at Neiman Marcus without waiting for the faux leather to fade.

- The federal government just wishes they had tagged the assets at Los Alamos and other sensitive facilities. You can track assets with RFID by, for example, triggering an alarm to sound and a camera to take a picture when tagged assets pass through a doorway. RFID allows all these things and more to happen automatically.

Mandates, Womendates, Blind Dates — Forcing Efficiency

In June of 2003, when Linda Dillman, Chief Information Officer (CIO) for Wal-Mart, announced to the world that Wal-Mart would require all suppliers to put RFID tags on every case and pallet that entered a Wal-Mart distribution center or store, the technology world as we knew it changed forever. This was the first of several high-profile mandates that rocked the retail and technology world and catapulted a new industry to be coined "the next big thing."

What are the major mandates?

This section gives you a rundown of the major mandates that are driving RFID implementation.

Wal-Mart

The Wal-Mart mandate detailed a plan for its top 100 suppliers to ship certain RFID-tagged items to distribution centers and stores in and around Sanger, Texas, by January 2005. Wal-Mart encouraged and engaged many other suppliers to participate — 137 in all. From that portentous announcement in June 2003, the press, the privacy advocates, and the competition began to emerge. The analysts quickly began to claim that RFID will be much bigger than Y2K and that Wal-Mart will become Big Brother and track everything everywhere. Sensationalism in the press took every angle from market size to predictions of failure. But no matter what angle they took, it was clear that the first stone was cast.

The U.S. Department of Defense

In the late summer of 2003, rumors of high-level U.S. Department of Defense (DoD) personnel making regular trips to Bentonville, Arkansas, began circulating in the RFID community. Rumors turned to rumblings when the DoD's Office of Automatic Identification Technology (AIT) began meetings with the various branches looking for information about existing RFID programs, the use of contact memory buttons, and where bar codes might be replaced and optimized by passive RFID tags. Although DoD was also an early member of the Auto-ID Center, the DoD was clearly going to use Wal-Mart's research and development efforts and early momentum to bring its own mandate to the world.

The DoD has always been a technology innovator through such groups as the Defense Advanced Research Projects Agency (DARPA) and others, but the technology impact has been mostly within its own secluded world. Demanding an RFID mandate of their 40,000 suppliers seemed like an unprecedented move — a move which had the potential to dwarf the impact of Wal-Mart's announcement in the technology and supplier world and guarantee the future of a fledgling RFID industry.

That announcement came in October of 2003, when Michael Wynne, Acting Under Secretary of Defense for Acquisition, Technology, and Logistics, released a policy paper spelling out a passive RFID program for all 40,000 DoD suppliers. When details were finally released in July of 2004, the policy turned out to be a near carbon copy of Wal-Mart's mandate. Cases and pallets going into two DoD distribution facilities — Susquehanna, Pennsylvania, and San Joaquin, California — are required to have passive UHF RFID tags with an EPC number or specific military number embedded on the tag.

Target

At about the same time the DoD announcement came out, another one of the most successful retailers in the United States, Target Corporation, announced its plans to keep up with Wal-Mart and require its suppliers to adopt RFID as well. Details of Target's mandate came out in August 2004, when the company called many of its suppliers to a meeting in the Minneapolis headquarters. The company took an intelligent approach to dealing with suppliers by making its mandate specific to a distribution center in Tyler, Texas. Target was also looking for suppliers that were already underway with Wal-Mart to participate in its early pilot, scheduled for a handful of suppliers in January 2005. The top suppliers to Target have until June 2005 to become compliant, allowing Target to stay a close follower to Wal-Mart, while learning from many of Wal-Mart's early mistakes.

Other mandates

Other mandates came along from the grocery store chain Albertsons, European companies Metro AG and Tesco, and (in a significant validation for the consumer products world) electronics superstore Best Buy. With many common suppliers in every industry deploying RFID, it is only a matter of time before other industry powerhouses like Home Depot, Lowes, Staples, and others follow suit.

Responding to the mandates

Mandates are similar to blind dates for many suppliers: The retailers say that RFID could be the perfect match, and that they're committed to seeing it through, but most of the suppliers haven't a clue what the outcome will be. As I write this book, suppliers have shown a range of responses:

- **Love at first site:** Some suppliers are already planning to adopt RFID deeply into their enterprise. Many industry pioneers have taken this approach. Gillette, Kimberly Clark, Procter & Gamble, Orco Construction

Supply, GTSI, and others have moved aggressively to gain a competitive advantage by incorporating RFID fully into their systems. These are the folks who are going to get an early — and potentially insurmountable — strategic advantage from the technology, in much the same way as FedEx crushed the U.S. Postal Service in overnight delivery by incorporating supply-chain optimization and technology into a delivery service. The Postal Service has never recovered. The companies investing heavily and working through the learning curve quickly have the potential to leave their competitors in the dust.

✔ **The cautious approach:** These suppliers are doing the minimal amount to get by until they discover more about the technology. This is a risk-aversion approach that doesn't lead to a big strategic advantage, but it also enables these companies to learn about the technology a bite at a time and not make any big mistakes in implementation — lower risk and lower reward.

✔ **The naysayers:** A small percentage of suppliers are doing nothing and will accept whatever penalties companies like Wal-Mart assess to noncompliant suppliers. These are the folks who, if they are in a competitive industry, are most at risk. Remember Eastern Airlines, and Digital Equipment Company? All once-successful companies that died because they failed to innovate. RFID represents a classic case of innovation advantage for early adopters and margin-eroding competitive pressure for naysayers.

Many folks may see a mandate as a powerful customer forcing new technology on a powerless client, and in some cases that is certainly the truth. The DoD, however, is a notable exception. According to analysts within the DoD's AIT group, the average payment cycle for a DoD supplier is 45 days from DoD receiving a shipment to a check being sent out to the supplier. With RFID-enabled shipments, DoD is committed to getting the payment down to 72 *hours.* The $60,000 question is when that efficiency will be in the system. My guess is that payment cycles will approach less than a week within four years.

Calling All Physicists! Calling All Physicists!

Over the past ten years, enough graduates have matriculated with a degree in physics to fill a few sets of New York City subway cars. Compare this with the number who have graduated with degrees in Engineering or Business Administration, which could fill up the entire island of Manhattan.

Why should you care about what Junior decided to study once he was out of high school? After all, the tuition is the same for basket weaving or applied physics, right? You need to know this because a jungle full of 800-pound gorillas in blue, monogrammed, smiley-face-adorned smocks are insisting that you need to use a technology you hadn't even heard of a year ago. The bottom line is that you're going to need help. You need a physicist.

Finding a physics expert

The marketplace dynamics of RFID are starkly different from the Internet, the word processor, the telephone, and other disruptive technologies of recent memory. In most other instances, invention, understanding, experimentation, and eventually adoption flowed naturally. Not so with RFID. Tens of thousands of enterprises are being forced to go from oblivion to adoption. This accelerated implementation creates a tremendous opportunity for the handful of folks out there who understand and can work with radio frequency technology. However, much like the carpetbagging that went on after the Civil War, it has opened the door for opportunists to try for a quick buck. And without many RFID experts in the world, you need the ability to distinguish the true experts from those who claim to be.

When you look for an expert to help with an RFID deployment, you can easily vet out the technology charlatans by having a little bit of knowledge and knowing the right questions to ask.

Because you're smarter than the average bear and bought *RFID For Dummies,* you'll at least know what you're in for and will eventually be able to choose a partner who can provide you accurate information and accurate help. Alternatively, brave warrior of RFID, I arm you with enough information to take on this mighty task yourself. Either way, to get you started, you need to understand something about the physics yourself.

The basic physics of RFID

In essence, an RFID system is just a reader and a tag communicating over the air at a certain frequency, like any other radio communication. The readers, antennas, tags, and frequency make up the basics of an RFID system, and the following sections give you an overview of how they work. Understanding some of the nuances behind the system as your company wades into the choppy waters of RFID can be the difference between making a multimillion-dollar mistake and being the CEO's new golfing buddy.

RFID readers

An RFID reader is really a radio, just like the one you have in your car, except that an RFID reader picks up analog signals, not hip-hop. The reader produces electricity that runs down a cable at a particular rate. That electricity eventually hits a piece of metal on the antenna, which radiates the same signal rate out in space at a certain frequency and wavelength.

The reader not only generates the signal that goes out through the antenna into space, but also *listens* for a response from the tag. The RFID reader is like a high-tech Morse code machine, but instead of the dots and dashes the Lone Ranger might have listened in on, the RFID reader transmits and receives analog waves and then turns them into a string of zeros and ones, bits of digital information.

Each reader is connected to one or more antennas. The three components, the reader, and the antenna are shown in Figure 1-1 and Figure 1-2. Figure 1-1 shows an Alien reader with Alien Class 1 tags (more on tag classes in Chapter 2) and Figure 1-2 shows a Matrics/Symbol reader with antenna and Class 0 tags. To put their size in perspective, the grid is made up of 12-x-12-inch squares. The antennas are a science all their own (see Chapters 4 and 5 for more details), but the important thing to know is that the reader creates the electromagnetic signal and the antenna broadcasts it into a specific interrogation zone. The interrogation zone is a radio frequency field that can be thought of as a giant bubble coming off of the antenna.

The tag

If the reader transmits a signal out into space (and space can be the distance from one side of a dock door to the other), what is out there transmitting back? The answer of course is the tag.

An RFID tag is made up of two basic parts: the chip, or integrated circuit, and the antenna. The chip is a tiny computer that stores a series of numbers unique to that chip. The chip also has the logic to tell itself what to do when it is in front of a reader. The antenna enables the chip to receive power and communicate, enabling the RFID tag to exchange data with the reader.

Some tags are *active tags* because a battery powers their communication. Most of the tags produced today are *passive tags*. This means that the only time they communicate is when they are in the close presence of a reader. Being in the presence of a reader means that they are sitting in an electromagnetic field. When a passive tag enters an electric or magnetic field, the tag draws enough energy from that field to power itself and broadcast its information.

The type of communication that allows this exchange to happen is called *backscatter*. The reader sends out an electromagnetic wave at one specific frequency. That wave hits the RFID tag, and the tag then "scatters back" a wave at a different frequency with the chip's information encoded in those backscatter waves. I explain how tags work with readers in more detail in Chapter 5.

Frequency

Both the tags and the readers operate over a specific frequency. Think of them as what they really are: radios that have their own very specific stations on which they can talk and listen. So in a way, the tags are tuned into the readers, just as your car radio is tuned into that hip-hop station.

Figure 1-1:
An Alien
reader, a
single
antenna,
and three
types of
Alien Class
1 tags.

Figure 1-2:
A Matrics/
Symbol
reader, a
single
antenna,
and two
types of
Matrics
Class 0 tags.

When will tag cost please the boss?

As I write this book, tags cost anywhere from $.22 to $1.20 each for passive tags, depending on volume, manufacturer, and special design functions for hard-to-tag items like metal or liquid products. Many people in the consumer packaged goods (CPG) industry have said that the "magic price" for tags is under $.05 each.

Many highly innovative companies are addressing this cost problem by pioneering production systems, experimenting with low-cost adhesives, and using conductive ink for antennas. Given the volume of potential applications (12 billion items in the pharmaceutical industry alone) and accelerating innovation, I advise clients who will buy in significant volumes that they should plan for a $.05–.075 tag by the end of 2007. Look for some large Asian manufacturers to bring tags to market in 2005 and 2006, adding to the increase in competition and fueling price pressure for that cheaper tag.

The majority of RFID being used in the supply chain world uses the ultrahigh-frequency band, or UHF. In the United States, this is referred to as the 915 megahertz (MHz) band. Although it is actually the 902–928 MHz range, 915 just happens to be the center. In Europe and Asia, this range is slightly different. Some applications, such as pharmaceuticals and asset tracking, use high frequency, or HF, which is at 13.56 MHz. Chapter 3 explains frequencies in more detail.

Finding Success with Four Ps in a Pod

I can enlighten you on all you need to know for an RFID deployment with the Four Ps. By the Four Ps, I don't mean that intoxicating Irish pub in Washington, D.C. I mean the four principal stages of an RFID deployment: Planning, Physics, Pilot, and Production.

The Four Ps encompass the key stages of an RFID network deployment. Figure 1-3 shows how they tie together in an evolutionary process of assessment, deployment, and scalability. The following sections explain the importance of each P in more detail.

Planning

Planning is the most important step in any complex undertaking. An RFID system is no different than a military operation; only the stakes are different. If you're playing the role of Captain RFID in your organization, the best thing you can do is plan properly.

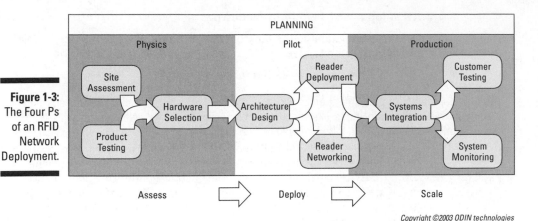

Figure 1-3:
The Four Ps
of an RFID
Network
Deployment.

The Planning stage ideally takes place over several months to make sure your organization has considered all the potential areas of impact, had time to get up to speed on the technology, and appropriately budgeted for the future. If you are being forced to comply by the government or a by large retailer, you do not have that luxury. If you do not have the time to do a full-blown planning session, use the following guidelines as must-haves for moving forward successfully. If you do have the time, use Chapters 3 and 16 as the foundation for your long-term planning cycle.

The critical planning steps for your RFID deployment are

1. **Create a global RFID policy.**

   Creating the global RFID policy requires a lot of research so that you understand all the available options in technology, business processes, and costs. This policy step addresses how to roll out the plan throughout your organization, what frequencies to use, data synchronization methods, and so on. If you are under a mandate, you might have many of these issues decided for you by someone else. Essentially, your global RFID policy will set the basis for how you need to move forward and help everyone in your organization understand what is about to happen with RFID.

   Spend as much time as necessary in setting the RFID policy. If you don't nail down your RFID policy well, you won't be overly successful with the steps that follow. Remember what Roger Staubach, a former Navy midshipman, once said: "Spectacular achievements are always preceded by unspectacular preparation."

2. **Execute an application analysis.**

   An application analysis covers the rationale and reasons for the RFID deployment and how RFID will be used. This includes very specific understanding of how RFID fits within your business processes. See Chapter 3 for more about assessing business processes.

3. **Develop a cost/benefit breakdown.**

   You need to examine the tangible and intangible ROI (return on investment). Chapter 17 gives a quick and dirty example of how to do that.

4. **Develop an implementation model (timeline).**

   With a new war chest of knowledge, you need to put together a project timeline (working back from any mandates you might be under) and investigate RFID vendors and consultants.

   Keep in mind that most RFID hardware vendors are not ramped up for high-volume production. This means order lead times can be anywhere from three to six weeks. Many people fail to consider this and end up being delayed several weeks. Make sure you incorporate equipment order timing into project planning timelines to stay on schedule. Chapter 12 offers more details on project planning.

5. **Design a deployment plan.**

   Basically, this step involves going through each step in the implementation model and assigning roles and responsibilities, seeing what parts are dependent on successful completion of other parts, and understanding the scope of the entire project. Having a timeline and some outside expertise on board will help you move toward a comprehensive, straightforward pilot that will serve as the foundation for a widely-deployed RFID network. See Chapter 17 for details on how the RFID plan fits in with your strategic plan.

6. **Manage the change and potential impact on the enterprise.**

   Finally, as with all good projects and consistent with the popular tenets of Six Sigma management principles, you need to audit the result by seeing how your deployment of the technology compared with what was being used before (usually bar codes) and ensure the survivability of the change by making sure the organization has mechanisms to prevent workers from avoiding or faking the use of the technology.

Physics

The second P is the Physics component. Certain laws of physics — no matter whom you know in the RF Police — just can't be bypassed. Those laws of physics are important because they affect the products you tag and the facilities where you set up readers. The three areas in which physics most come into play are

 ✔ **Full Faraday Cycle Analysis to understand the environment:** The Full Faraday Cycle Analysis, named after the famous physicist of the 1800s, Michael Faraday, is made up of two primary components. First is a time-based analysis of ambient electromagnetic noise (AEN), and second is RF path loss contour mapping (PLCM). You can find out how to execute both of these functions in Chapter 7. The goal is to see all the invisible electronic, magnetic, and radio waves that flow throughout a location and then properly design an RFID network to live within that environment.

✔ **Product or SKU testing for tag selection and placement:** This step involves properly testing your products for an RF signature. Many people refer to this as *SKU testing for RFID compatibility.* In a vacuum, the reader and its antenna combine to make a perfectly shaped RF field. Put an object, like a case of SPAM, in the middle of that field and that perfectly shaped RF field becomes distorted beyond recognition. Why? Because RF waves, like light waves, can be reflected and absorbed. Metal reflects RF waves, and liquids absorb them. Knowing this, you can imagine how an RF calamity might ensue in an interrogation zone if you try to tag a case of SPAM, a highly liquid foodstuff in a metallic can. To avoid this calamity, see Chapter 8, which goes over a sound scientific methodology to find the right tag and placement for your products.

✔ **Selection of the RFID hardware based on scientific testing:** Buying an "RFID in a box" or a "slap and ship portal" is a big mistake. Although these solutions look attractive on the surface, they can turn into a maintenance and support nightmare, and often end up being completely written off as organizations move to a full RFID network. The physics and planning should be done with the end in mind — where do you expect or want your RFID network to be in three to five years? If you are planning for ten dock doors, design for that and source a solution *now* that is optimal for the long term, even if you're setting up only one dock door today.

To design with the end in mind, you need to do scientific testing. After you understand how your products behave in an RF field and what the specific requirements of your environment are, you can set up a lab to help you discover what the best readers and antennas are. Colvin Ryan, the world-famous steeplechase jockey, is famous for saying, "No matter what place the horse is in over the first two fences, the only thing that matters in the end is who gets the girl." That is a prime example of working with the end in mind.

I remember one client whose software vendor sold them a print-and-apply solution and readers before any of the physics testing was done. The client then discovered that the tags read 10–15 percent of the time at most and that the readers didn't have the communication capabilities to fit well into the existing infrastructure. Then they went through the proper testing and hardware selection and are now at a 100-percent read rate. But they're left with several thousand dollars' worth of high-tech paperweights.

Pilot

If you've been following the RFID buzz for the past year or two, you might think you were on the set of the movie *Top Gun* with the number of times you've heard the word *pilot* bantered about. The truth is, there is so much to learn about RFID that companies are trying to get away with as little initial impact as possible. Many are limiting the commotion by starting out with a one- or two-location pilot, or a trial system.

The bad news is that pilot costs can range from $50,000 to $1,000,000 depending on the scope and requirements. The good news is that, when done correctly, a pilot program can save you hundreds of thousands of dollars as your company moves toward full deployment. And you *will* eventually be deploying an RFID network. Think of the pilot as an initial deposit in a high-yield 401(k) — the earlier you start it, the more benefit you get out of it in the long run. (Sorry, I know this isn't *Financial Planning For Dummies,* but that recessive MBA gene flexes its helix every now and again.)

In essence, the pilot provides a solid road map for production but has a more limited scope. Following the Four Ps process, the Pilot stage becomes a pragmatic step toward true understanding of RFID.

The pilot is about deploying and testing the RFID network in your environment. To get a better sense of what a pilot involves, see Table 1-1, which outlines the basic phases of an RFID pilot.

Table 1-1	Phases of an RFID Pilot	
Phase	*Percent of Total Pilot Timeframe*	*Key Tasks*
Planning	40 percent	Designing a single RFID interrogation zone to work in concert with business processes and systems Testing for proper hardware choice; the better the planning, the fewer the changes after deployment
Setup and installation	30 percent	Putting together the hardware, configuring it, integrating it with existing systems, and then training users
Testing and redesign	30 percent	Evaluating the performance of the design and process and making modifications to increase performance

Think of the pilot deployment as the first node in an overall system architecture that may take years to develop completely. Pilots provide a road map for production, but have more limited scope and a longer redesign process. After the system is up and actually collecting data, you can expect several reiterations of design and modifications to the process. That redesign process allows you to expand the system as you're ready; it also helps you understand that the RFID network is a living thing, evolving as business processes change and become optimized. As shown in Figure 1-3, the pilot logically morphs into the Production phase and scaling up the network.

Production

The first three Ps might seem like a sprint to get yourself ready with this new technology, but the last P, Production, is the methodical scaling up of a well-designed system. It's the steady pace of a marathoner who knows exactly what his splits should be at every mile to get to the finish line.

The key difference between the pilot and the production systems is that the network grows exponentially in complexity as readers are added and more data is captured. As scary as this might sound, if the Planning and Pilot stages were done with the end goal in mind, growth should come smoothly and relatively painlessly. Scaling up an RFID network is similar to the pilot process; you add nodes to a previously designed network and focus on small design modifications to manage any unplanned events.

In addition, when you reach the Production phase, you're ready to add the following tasks into the mix of your RFID network:

- **Managing the health and performance of the network:** This is the most complex challenge of production and involves making sure that the readers are performing optimally and stay correctly configured. Detecting anomalous behavior before it leads to catastrophic failure is the key. Only a couple of options today address this need, and they are covered in Chapter 14. One thing is very clear, however: Traditional network management systems like Tivoli, Unicenter, and OpenView are poorly suited for management and monitoring of a complex RFID network because they can't understand the multifaceted physics components that are at the root of an RFID network's performance.

- **Integrating your RFID data into existing systems:** This is the timeliest issue. An RFID network will produce much more data in real time than your current system (because items are serialized). This is very different from what most core business applications are used to. Many are designed to deal with bar code data coming in at regular intervals in a batched mode. Fortunately, the major enterprise resource planning (ERP), warehouse management (WMS), and inventory software vendors are designing and building new additions to their existing applications specifically for RFID. This will ease the integration burden and help enterprises leverage the intelligence gleaned from real-time serialized data. Already, companies like SAP, Manugistics, Oracle, and others have built RFID middleware and modules that their existing clients will be able to benefit from. See Chapter 10 for more details.

✔ **Testing your system with outside partners:** Just like the force of an army's battalion is made up of many individual soldiers, the power of RFID is unleashed when a multiplicity of single nodes are bonded together sharing real-time, specific data. After you have data populating your critical business applications and are confident your RFID network and infrastructure are performing well, you can start to test with selected suppliers and customers. The value of this information is stunning:

  • *For companies concerned with inventory management,* incorporating both upstream and downstream partners provides a level of in-transit visibility that allows radical changes in your inventory management process and, most importantly, reduces necessary capital tied up in the inventory cycle.

  • *For companies focused on asset tracking and security,* incorporating the new RFID data with back-end applications allows chain-of-custody or pedigree information and specific association with people, plants, and distributors that has never been available.

✔ **Educating the users:** Training is critical to ensure front-line adoption and proper usage of the systems. The complexity of performance and the invisible nature of RF make for a unique combination for the worker in the field. Warehouse and system staff need to understand what affects the success of a reader network and how to recognize some of the basic issues. Behaviors they may not think twice about today may need to be modified. For instance, if a worker decides to unplug a reader to use the outlet, he needs to know that the custom configuration on most of today's readers will be lost, and when that reader is plugged back in, that the con-figuration is set back to the factory default. Or if a forklift is parked in a reader's interrogation zone, users need to know that the success of tag reads is likely to be altered. Performance and business process issues can be designed into the network to a certain extent with visible light or sound queues, but many of the relevant issues will need to be addressed with specific training. Chapters 14 and 15 discuss training for your pilot and production deployments in more detail.

A Ride in the Time Machine

This book was written in 2004 and released in early 2005. So what will things look like five or ten years from now? As I mentioned earlier, the $.05 tag will be a reality, but more importantly, RFID technology and a global protocol will enable a world we couldn't even have imagined at the turn of this century. In less than five years, we will witness a $25 RFID reader and all the technology and digital signal processing on a single chip. RFID readers will come in two flavors: (1) cheap, dumb readers that only read tags and send the data up to a central collection point, which filters and smoothes the data for analysis; and (2) more expensive, higher-processing, smart RFID readers that can perform intelligent operations beyond simple communication.

The cheap and small readers will enable a convergence of parallel technologies you may have already heard of:

- **Mesh networks:** Items that communicate and self-configure every time a new node is recognized or removed.

- **Grid computing:** The ability to co-opt computing power like a utility when an application needs more horsepower.

- **Dust motes:** Tiny sensor networks that can do everything from predict disasters like tsunamis to recognize chemical warfare, and can be deployed by dropping them from a plane by the thousands, like crop dusters.

- **Sensors:** To monitor everything from temperature to vibration to nuclear levels attached to this networked world.

Many people's vision of an internet of smart objects will be realized as all of these technologies unite in a manner that is pure machine-to-machine communication. An object embedded with an RFID tag or some derivative will enter the presence of other objects that are similarly enabled and be instantly recognized. Each object will have enough data to configure itself into the geographical network in which it resides. Information about everything from temperature and movement to cost and ownership will be distributed in these complex systems. Most importantly, it will all happen wirelessly, with a limited number of data standards, such as the EPC protocol, ISO standards, and unlicensed bandwidth.

This intelligent, wireless, machine-to-machine communication will grow at a cost of strict regulatory compliance related to our privacy and freedom if we let one seed germinate that has already been planted — ignorance. Without clearly understanding the impact and application of disruptive technologies like RFID, some people will have a knee-jerk reaction that our privacy is at stake. In today's world, and the world a decade from now, education and understanding will be the best protection against overbearing federal regulations and alarmist articles in popular press. See Chapter 2 for more about privacy concerns.

George Jetson never had it as good as we will a decade from now: When you wake up in the morning, your armoire will notify you of your perfect wardrobe based not only on fashion coordination, but also on what's on your calendar for the day. Then you'll get in your car and head back to sleep because it will use sensors and GPS to drive you to the office. If a tire gets low on air, it will route the car to the service station for a quick fix and get you back on your way. If the interstate is backed up, your car will route you to a different path and get you there on time. Then, when you stop at the store on the way home, you'll simply fill your cart up with items and walk out the front door, and a sign will let you know that you've just purchased $38.76 worth of Pop-Tarts, toothpaste, SPAM, and Cheez Whiz. When you get home, your refrigerator will display a warning light that your arteries are going to be blocked quicker than brushing your Newfoundland in the tub will block your drain if you don't change your diet. If it does get that far, we may want Jane Jetson to stop this crazy thing called technology. But, needless to say, the future could be a very cool place that is wildly efficient, thanks to RFID.

Chapter 2

Auto-ID Technologies: Why RFID Is King of the Hill

In This Chapter

▶ Investigating automatic identification technologies

▶ Comparing various numbering schemes for Auto-ID

▶ Understanding privacy and standards

. . . .

*T*he world of automatic identification technology (referred to as either *AIT* or *Auto-ID*) has steadily grown over the past half-century into what it is today — an indispensable part of our everyday life. The bar code has been the sovereign monarch, the foundation and foothold for Auto-ID technologies, for the past three decades. Look on the back of this book, on your computer, under the hood of your Ferrari — and yes, even tattooed on the neck of that kid down the street with all the piercings: Bar codes are everywhere. And if you look closely enough, you might just notice a few gray lines in with the usual black ones: The bar code has entered its thirties and is showing some subtle signs of age.

Although RFID is the new heir apparent, it isn't replacing every bar code or other Auto-ID technology any time soon. Each Auto-ID technology has its strengths and weaknesses. Even though the bar code is getting a little gray around the temples (like me), it still has plenty of use to the industrial world — hopefully, also like me.

As you look at deploying Auto-ID technologies in your enterprise, understanding what the key Auto-ID technologies (RFID, bar codes, and contact memory buttons) can and can't do is important. Knowing the strengths, weaknesses, costs, and issues of each one will help you craft a strategy that incorporates the best possible options. Having a good grasp of the standards surrounding the newest of the technologies, RFID, will help you glance into the future and plan for adoption time frames, interoperability issues, and data synchronization problems.

Planning an Auto-ID Strategy for the Times

In this section, I cover the key features of the bar code and other Auto-ID technologies so that you can see how newer ones can take advantage of developments in technology infrastructure.

The three principal types of Auto-ID technology that I cover are bar codes, contact memory buttons, and radio frequency identification (RFID). All three technologies have a viable place in the global marketplace today. The distinct technology differences show why there are very specific applications for each of the three. After you understand the key features of Auto-ID technologies and how the different options stack up, you can begin to think of strategies for using them in your business.

Whenever you compare different items of the same ilk — cars, ski goggles, shoe polish, whatever — it helps to have a few criteria to work with. The following list gives you some idea of the criteria to compare Auto-ID technologies, which can help you figure out which ones best fit your business needs:

- ✔ **Modification of data:** The ability to change the data on the tag or to write data to the tag.

- ✔ **Security of data:** The ability to encrypt the data on the tag.

- ✔ **Amount of data:** The amount of useful data the tag can store.

- ✔ **Costs:** In addition to the obvious — how much each one costs — also remember to consider the costs of ancillary equipment you need to work with a technology.

- ✔ **Standards:** Whether there is a set open standard that many manufacturers and users have adopted, or whether the technology is proprietary to one manufacturer (remember VHS versus Betamax?).

- ✔ **Life span:** How long the tag remains readable. Some tags enable you to read their data indefinitely, whereas others have a shelf life.

- ✔ **Reading distance:** Whether the tag requires line of sight to be read and how far away can it pick up a signal.

- ✔ **Number that can be read at a time:** You read a bar code or contact memory button only one at a time; other technologies enable you to read multiple tags at a time.

- ✔ **Potential interference:** What can keep the tag from properly being read.

Now take a look at how the three primary Auto-ID technologies do in each category. Although the RFID technologies and standards are rapidly evolving, Table 2-1 shows a quick summary of the three technologies today. I explain the key features and how they differ for each one in the sections that follow.

Table 2-1	Comparing the Primary Auto-ID Technologies			
	Bar codes	*Contact memory*	*Passive RFID*	*Active RFID*
Modification of data	Unmodifiable	Modifiable	Modifiable	Modifiable
Security of data	Minimal security	Highly secure	Ranges from minimal to highly secure	Highly secure
Amount of data	Linear bar codes can hold 8–30 characters; other 2-D bar codes hold up to 7,200 numbers	Up to 8MB	Up to 64KB	Up to 8MB
Costs	Low (pennies or fraction of a penny per item)	High (more than $1 per item)	Medium (less than 25 cents per item)	Very high ($10–$100 per tag)
Standards	Stable and agreed	Proprietary; no standard	Evolving to an agreed standard	Proprietary and evolving open standards
Life span	Short unless laser-etched into metal	Long	Indefinite	3–5-year battery life
Reading distance	Line of sight (3–5 feet)	Contact required	No contact or line of sight required; distance up to 50 feet	No contact or line of sight; distance up to 100 meters and beyond
Potential interference	Optical barriers such as dirt or objects placed between tag and reader	Contact blockage	Environments or fields that affect transmission of radio frequency	Limited barriers since the broadcast signal from the tag is strong

Comparing the major players in Auto-ID: Bar codes, contact memory, and RFID

The granddaddy of the Auto-ID technologies is the bar code. In fact, the very first item to be bar coded, a pack of Wrigley's chewing gum, is now on display at the Smithsonian Institution, reminding future generations of the pivotal role the technology played. However, the bar code has limitations that don't take advantage of the technical infrastructure available today. The following sections explain what the different technologies are and break down the details of what bar codes have to offer compared with contact memory buttons and, of course, RFID.

Bar codes

The problem with the bar code is that the maximum throughput in any bar code system is one: that is, you can scan only one object at a time. In addition, because a limited amount of data is stored in a small form factor, the bar code doesn't have enough room for a unique serial number, expiration date, or other pertinent information. Lastly, the bar code reader has to be able to "see" the bar code marking to read it. For example, if a bar coded item is wrapped, packaged in a container, kept under a sheet or cover, or has somehow gotten dirty, dusty, or marked, the bar code can't be read.

Because of these limitations, most bar code innovations in the past few years have focused on data-capture and data-transmission devices to make bar codes more useful and to help them keep up with faster computing power and better network connectivity. This section explains the different bar code systems available — the old linear bar code, the stacked bar code, and matrix symbols — and clarifies how they stack up.

Linear bar codes are the most widely used Auto-ID system. They can be found on everything from cans of soda to rental cars. They are formed by printing a series of alternating dark and light (usually white) bars of varying width. These patterns have very specific meanings and representations. The other component of an Auto-ID solution, the reader or scanner, is of course a key part of a linear bar code system. Many types of scanners can read linear bar codes. Fixed-location scanners can be used to read linear bar codes without significant operator intervention if there is a method to ensure that the label faces the scanner. In terms of the criteria discussed earlier, linear bar codes offer the following:

- **Modification of data:** After a bar code is printed, it's done. You can't change the orientation of the markings after the symbol has been printed or etched.

- **Security of data:** Linear bar codes are widely adopted, and the standards are well known; however, they are not encrypted for security.

- **Amount of data:** Linear bar codes can have up to 30 characters of data.

✔ **Costs:** The cost can be a fraction of a penny or several cents if the bar code is etched into an item.

✔ **Standards:** One of the shortcomings of bar code technology has been the lack of a true universal protocol; the good news is that many of these standards are quite stable and are adopted by many end users. This is illustrated by the fact that over 200 types of bar code schemes *(symbologies)* are in use today. Realistically, however, only four symbologies (UPC/EAN, Interleaved 2-of-5, Code 39, and Code 128) are in common use, and all are covered by International Organization for Standardization (ISO) standards.

✔ **Life span:** Life span is fairly low because they are usually printed. However, if they are etched, they can last a very long time.

✔ **Reading distance:** Linear bar codes require line of sight to be read and have a range of a few feet.

✔ **Number that can be read at a time:** Only one item can be scanned at a time.

✔ **Potential interference:** Linear bar codes become unreadable when vertical damage occurs. Such damage occurs when a black bar is completely eliminated or altered or when a white bar is filled in. In the event of vertical damage to the symbol, there is typically no possibility of recovering the data. Only one bar code symbology (93*i*) contains erasure and error-correction capabilities. The symbol also becomes unreadable if obscured by dirt or other contaminants or when severely abraded. In addition to the bar code being susceptible to dirt and dust, the readers also cease to function if dirt, dust, or other foreign objects obstruct the lens.

Another type of bar code is the *stacked bar code* (also called a *2-D bar code*). From a technology perspective, a stacked bar code comprises multiple rows of very short linear bar codes, arranged in a specific manner to ensure correct decoding. Although several stacked bar code symbologies are available, only one is commonly used: PDF 417. The stacked bar code is very similar to the linear bar code, with the exception of the following key differences:

✔ **Security of data:** Because they lack the vertical redundancy of linear bar codes, stacked bar code symbologies employ a specification called *Reed-Solomon erasure and error correction,* which allows part of the tag to be destroyed while retaining all the original information. Data compaction schemes as well as encryption help to increase data capacity and enhance data security. Because it is a line-of-sight technology that carries more data than a simple linear bar code, additional security concerns exist. For instance, a PDF bar code can be photocopied, scanned, or faxed and subsequently read, making counterfeiting and theft very simple — a continuing problem with the bar codes used on tickets for sporting events.

- ✔ **Amount of data:** The stacked bar code is the only bar code on which a significant amount of storage can be added right to the tag. Stacked bar code symbols can contain more data than linear bar code symbols — up to a full kilobyte.

- ✔ **Cost:** Stacked bar code symbols are less readily available from third-party vendors, but despite a significant amount of competition, they still share the ability to be printed or etched and therefore are very low-cost.

- ✔ **Standards:** PDF 417 is an ISO standard. PDF here stand for *portable data file* (not the Adobe portable document format). This symbology addresses many of the limitations of linear bar codes and has been the only genuine innovation in tag design in recent years.

- ✔ **Potential interference:** Although they are more tolerant of localized damage than linear bar codes, significant amounts of obscuring material or abrasion can still render them unreadable in spite of their error-correction capabilities

Matrix symbols are yet a third type of bar code. They're composed of discrete modules (typically round or square) arranged in a grid pattern. In the United States, the most widely known examples of a matrix symbol are the codes that the U.S. Postal Service prints on letters and postcards in order to sort the mail.

Matrix symbols share many characteristics with the linear bar code, but they do have some unique traits that make they better suited for specific applications:

- ✔ **Security and amount of data:** In these areas, matrix symbols have the same capabilities as stacked bar code symbologies and are roughly equivalent in data capacity and error correction.

- ✔ **Costs:** Matrix symbols can be read only with two-dimensional array (charge-coupled device [CCD] or complementary metal oxide semi-conductor [CMOS]) readers, which are more expensive than standard bar code readers.

- ✔ **Standards:** A number of matrix symbologies are available, but only three are in common use: Data Matrix, QR Code, and MaxiCode. Data Matrix symbology is covered by an ISO standard. ISO approval of QR Code is pending. Only United Parcel Service (UPS) uses MaxiCode. Aztec and Mesa Code are two other less commonly used matrix symbologies that are undergoing ISO standardization.

Matrix symbols for harsh environments are infrequently available from third-party vendors, although there is support from some direct-part-marking equipment vendors.

Matrix symbols are more tolerant of printing irregularities than width-based symbologies such as linear and stacked bar codes. Additionally, direct marked symbols have very low contrast between "marked" and "unmarked" areas.

Contact memory buttons

Contact memory buttons have also been around for nearly a generation. They are a specific type of Auto-ID that requires a wand to make physical contact with a button tag to read the data on the tag. Each button tag is about the size of a quarter. Given the limited adoption of contact memory button technology, comparatively little investment and innovation is occurring in this arena.

Because contact memory will never be a widespread Auto-ID solution, a key concern surrounding this technology is that the three major contact memory button solutions in use today are proprietary systems. If those solutions are discontinued, finding a replacement may prove difficult. But as you can see from some of the key attributes, contact memory does have some distinct advantages.

- ✔ **Modification of data:** Contact memory buttons can be written to and read many times. They are robust because they can withstand vibration and harsh environments and still be read.

- ✔ **Security of data:** Contact memory buttons can have their data encrypted.

- ✔ **Amount of data:** Data storage can be up to 8MB.

- ✔ **Costs:** Start at just over $1.

- ✔ **Standards:** There is no universally accepted standard; contact memory buttons are proprietary technologies.

- ✔ **Life span:** The physical contact required for communication with the reader limits the usable life of that reader.

- ✔ **Reading distance:** Because the tag reader has to come in physical contact with the button tag, the reading distance is essentially zero.

- ✔ **Number that can be read at a time:** You can read these only one at a time.

- ✔ **Potential interference:** The physical contact required also limits the efficiency with which the contact memory button can be read.

RFID

An RFID solution uses a radio frequency (RF) signal to broadcast the data captured and maintained in an RFID chip. An RFID system is composed of three components: a programmable transponder or tag, a reader (with an antenna), and a host. Figure 2-1 shows the basics of how an RFID system works.

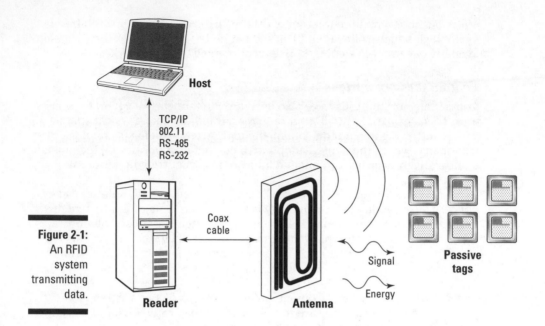

Figure 2-1:
An RFID
system
transmitting
data.

Much of the criteria for RFID systems depend on the type of tag that you use. Tags can be active, passive, or semi-passive. Here's an overview of the different types:

- **An active tag** has its own battery power to contact the reader. Power from the battery is used to run the microchip's circuitry and to broadcast a signal to a reader. An active tag's onboard power source enables the tag to broadcast a signal out at great range by either constantly beaconing a signal or broadcasting only when the reader talks first. Some of the more powerful active tags can communicate up to 1 kilometer.

    Active tags are much larger and therefore can carry a lot more memory capability. Rather than simply having a unique serial number on the tag, like a passive tag, active tags often carry information such as the full contents of a container, its destination, and its origin. By carrying all the information on the tag, you can retrieve information instantly. For example, soldiers in battle usually can't look up a file associated with a tag on the Internet, so soldiers use hand-held units to scan containers with active tags to find out what's inside.

    Despite their cost, active tags have proven a significant return on investment (ROI) for many applications. Since the early 1990s, the DoD has put active tags, about the size of a cigarette carton, on containers to track both their contents and their whereabouts. As of 2003, *every* container that the DoD ships out of the United States has an active tag affixed to it. Certain types of active tags used in the railroad and shipping industries can integrate with an onboard Global Positioning System (GPS), cellular communication network, or satellite system to give exact whereabouts and provide constant communication back to a tracking program.

One of the reasons active tags have not grown to widescale adoption is the lack of an open global standard, although the DoD has helped to drive efforts toward creating interoperability with active tags.

✔ **A passive tag** does not require a battery. Rather, a passive tag derives its power from the electromagnetic field created by the signal from the RFID reader to respond to the reader with its information. Because the mandates generally require passive tags, I devote more space to explaining how they work in this book. See Chapter 5 for details on the physics of passive tags, and Chapter 8 for more about testing for the right tags.

✔ **Semi-passive tags** use a battery to run the chip's circuitry but communicate by drawing power from the reader's radio waves (like a passive tag). Because these tags have a battery, they're larger and more expensive than passive tags, but have greater communication ranges. Some active tags can also be made to monitor sensor inputs, such as the temperature or movement, even without being within an interrogation zone to power up the tag.

In addition, tags come in different classes and generations: Class 0, Class 1, and Gen 1, Gen 2, and so on (which, for the purposes of comparing RFID against the other Auto-ID technologies, determines whether you can modify the data).

Here's how the different criteria break down:

✔ **Modification of data:** The ability to modify data depends on the standard that you use. Using the electronic product code standard (EPC), the two classes are

- *Class 0 tags:* These are read-only, which means you have to use the number that the manufacturer writes on the tags.

- *Class 1 tags:* These are read/write tags, which means you can program whatever number you want on them (often called *commissioning* the tag) at your place of business and then read them an infinite number of times — write once, read many (WORM).

✔ **Security of data:** Depending on the class and generation of the RFID tag, they have the ability to be encrypted so that others with standard RFID readers cannot read the actual data on the tag. For more details about the security of data in RFID systems, see the section, "To EPC or Not to Be: Unraveling the Words, Words, Words of the Electronic Product Code," later in this chapter.

✔ **Amount of data:** Depending on the manufacturer, these tags can contain 64, 96, 128, 256, or 512 bits of information.

✔ **Costs:** Costs, which range in volume from 20–50 cents per tag, are falling fast.

✔ **Standards and regulations:** RFID systems use many different frequencies. International governing bodies — such as the Federal Communications Commission (FCC) in the United States and European Telecommunications Standards Institute (ETSI) in Europe — regulate these frequencies. Generally, the most common are low-frequency (LF, around 125 kHz), high-frequency (HF, 13.56 MHz), and ultrahigh frequency (UHF, 850–930 MHz). UHF RFID systems have been in commercial use only since the mid-1990s, and countries have not agreed on a single area of the UHF spectrum for RFID. Europe centers around 868 MHz for UHF (and as recently as November 2004, increased some of the available power and spectrum), and the United States centers around 915 MHz. Radio waves behave differently at different frequencies, so you have to choose the right frequency for the right application. I discuss frequency in detail in Chapter 3.

As far as standards go, the protocols that the tags and readers use to communicate can be described as the *air interface protocol*. The two primary ones that you hear about are the EPC standard and the ISO standard. Both EPC Class 0 and Class 1 tags and ISO 18000-6 tags operate in the 860–930 MHz range. A new standard called EPC Gen 2 is being developed that will also work in the 900 MHz range but will be a worldwide standard for data communication. I discuss these protocols in detail later in this chapter.

✔ **Life span:** Having no need for a battery makes the passive tag's life virtually unlimited. Active tags and semi passive tags last as long as their batteries. Refer to Table 2-1 for more details.

✔ **Size:** Passive tags range in size from Hitachi's mu-chip (μ-chip; about the size of a pin head) to the size of a letter envelope. Active tags range in size from the size of a piece of hard candy to about the size of a carton of cigarettes. The larger the size, the fewer items the tag can easily be affixed to. The smaller the size, the less the read distance.

✔ **Reading distance:** Passive tags communicate in ranges from a few millimeters (called the *near field*) all the way out to tens of meters. Active tags can communicate more than 100 meters. The big benefit, as I mention earlier, is that you don't have to see a tag to read it. Tags inside containers, behind walls, in briefcases, and so forth can still be read.

✔ **Number that can be read at a time:** A reader can read hundreds of tags nearly simultaneously.

✔ **Potential interference:** Various materials such as metals and liquids can interfere with passive tags. Active tags are less susceptible to interference but still can have issues inside metal containers.

MEMS, the kissing cousin of RFID

A MEMS device connects the physical world with the electronic world. MEMS not only identify a component or product (like an RFID tag), but also take measurements from or make changes in the physical world. MEMS devices come in several functional types:

✔ **Sensors:** *Sensors* detect a condition in the physical world and convert it to an electrical signal or data element (for example, a tire with a MEMS device embedded can report its pressure back to the car).

✔ **Actuators:** *Actuators* convert an electrical signal into an action in the physical environment, such as a garage door opener.

✔ **Combination:** Sensors and actuators can be combined in devices such as thermostats.

MEMS devices typically use RF transmissions to communicate information to other systems or devices.

Remember: RFID is used to identify items with a number and is therefore an open system. This means that it requires access to other systems to complete its function. MEMS devices, on the other hand, take measurements of the physical world and can do so in a closed-system function. In other words, MEMS can be entirely independent of other systems, or they may carry an identification number and thus become an open system. (***Note:*** At this point, MEMS identification numbers appear to be proprietary and not based on open standards.)

Two types of technologies use radio frequency as a broadcast mechanism: radio frequency identification (RFID) and micro-electro-mechanical systems (MEMS). Unfortunately, the market and internal users sometimes confuse these two technologies. In many cases, people use the term *RFID* to refer to both technologies. It's important to understand the two are different. See the sidebar, "MEMS, the kissing cousin of RFID," for details.

Crafting an Auto-ID strategy for your business (Or, why RFID is the wave of the future)

The preceding section gives you an overall idea of what the different Auto-ID technologies can and can't do. Maybe you've already begun to see how your Auto-ID strategy can evolve beyond the bar code. Regardless, this section digs a little deeper, connecting the specs to real-world practices.

Bar codes are cheaper and, in a few specific instances, perform better than other forms of Auto-ID. This fact supports the logical conclusion that a combination of bar code and RFID will become the optimal strategy for most enterprises. Bar codes still have the competitive edge in the following cases:

- **When you need to apply the ID directly to an asset, particularly when you need to etch the ID directly in metal:** Linear bar codes are widely used by commercial, industrial, and governmental agencies to tag everything from boots to data storage devices. Even weapons have been laser-etched with linear bar codes by the Department of Defense (DoD). However, Matrix symbols, particularly Data Matrix, are better suited than other bar code types for direct marking on items. Several current applications in which Data Matrix codes have been permanently marked on metal and plastic have been successful on the market.

- **When you need to identify items out in the field:** For example, the development of bar code readers that attach to cellular phones enables delivery personnel, repair technicians, and security staff to identify items on location. Although active RFID tags offer the same capability because they have onboard power, the active tags are large and expensive to implement. The use of bar code scanners is more cost effective in these cases, especially because identifying items (such as a package you're having shipped to your house) don't require the increased data storage space that active tags offer. The postal folks can get the job done with a simple matrix bar code. Like many other things in the RFID industry, however, this is evolving over time. Now there are companies offering portable RFID readers that even interoperate with mobile phones. Nokia has one such solution that you can check out here: (www.nokia.com/nokia/0,,55738,00.html)

Likewise, contact memory buttons have useful applications as well, such as

- **Reading moving parts that create a lot of radio noise and vibration:** Because RFID relies on radio frequency, an RFID tag can't transmit a signal successfully through significant radio noise. (You can find out more about noise in Chapter 7.) Contact memory buttons are not affected by radio noise because the communication is via contact, which make them an ideal workaround for these types of environments.

- **Tracking volatile chemicals:** A contact memory button is still the preferred type of Auto-ID for this purpose because it is less susceptible to corrosion and failure in harsh environments.

With those cases out of the way, you can now focus on the interesting part — how do you best harness the potential of RFID? The evolution of technology — particularly applications such as enterprise resource planning (ERP) software, warehouse management systems (WMS), and assess management systems (which can handle greater volumes of near-real-time data) — has highlighted the shortcomings of the bar code. With RFID, however, you can take advantage of advances in technological infrastructure, which opens up many possibilities for businesses:

- ✔ **Tracking high-value goods that need to be scanned over long ranges:** Active and semi-passive tags have the great benefits of not needing line of sight, working at long distances, and reading many at a time with great speed. That's why RFID is so well suited for tracking many items on high-speed conveyors, moving forklifts, railway cars on a track, or with tens of other containers in a yard. Active and semi-passive tags cost a dollar or more, however, making them too expensive to put on low-cost items.

- ✔ **Tracking many items that are fast-moving and not of great value in a supply chain or asset management:** Companies are implementing passive UHF tags, which cost less than 25 cents today in volumes of 1 million tags or more. Their read range isn't as far as with active tags — typically less than 20 feet, versus 100 feet or more for active tags — but they are far less expensive than active tags and can be disposed of with the product packaging. The ability to dispose of the tags is becoming an important issue due to privacy concerns. This is why Wal-Mart, Target, the DoD, and many others have required their suppliers to use this technology — so that they can have a live, real-time view of their supply chains.

- ✔ **Tracking data in real-time with *serialized* (each item has its own serial number) data:** It is clear that these applications can handle data in near-real-time instead of the traditional batch mode. *Batch mode* means that a large volume of data is sent all at once to be processed overnight or on a weekend.

- ✔ **Using machine-to-machine communication to make decisions and set actions:** RFID can be incorporated into sorting systems such that, if a reader scans a particular case, it can send a signal to the conveyor to sort that case to a specific area: The machines make the decisions, not humans. It is also clear that to truly eliminate human error and to increase speed, the communication needs to be between machines, not from human to machine.

- ✔ **Using serialized data:** Because each RFID tag has its own unique serial number — not just a product identifier like a bar code — significant information can be gleaned about the supply chain if each item has its own unique serial number that can be stored for an indefinite period of time.

The bar code didn't easily allow for any of these processes.

To EPC or Not to Be: Unraveling the Words, Words, Words of the Electronic Product Code

What the UPC is to bar codes, the EPC (electronic product code) is to RFID. When an RFID reader scans a tag on a case of toothpaste, the tag sends an EPC number to the inventory management system, letting it know exactly which case of toothpaste it just saw, where it was, and at what time it saw it. If you are one of the 60,000 companies under a mandate to implement RFID, you need to do so by using the EPC protocol. Thus, understanding it helps you get your arms around not just the technology of RFID, but also the standards of EPC.

Birth of a revolution: The MIT Auto-ID Laboratory

Like many technology revolutions, the Auto-ID development has come about primarily because of the DoD. Since the early 1990s, the DoD has been using active RFID to tag shipping containers moving throughout the DoD supply chain. The private sector soon realized that the DoD's massive supply chain was a great proxy for any commercial supply chain: they could use RFID to track cases and items, rather than just shipping containers, and make a huge impact on their operations.

However, the retailers needed to sort out major shortcomings in price, size, and interoperability. The people most interested in seeing RFID blossom — namely, large retailers and consumer packaged good (CPG) manufacturers, along with the DoD — funded MIT's Auto-ID Center in 1999. The Auto-ID Center's goal was to make RFID a viable technology in the supply chain.

The MIT Auto-ID Center team was energized by the leadership of Sanjay Sarma and David Brock and was augmented by Kevin Ashton (an executive on loan from Procter & Gamble). The team needed some big foreheads to figure out all the tough issues, so Sarma enticed one of his former grad students, Daniel Engels, to finish his Ph.D. at MIT and help solve issues that the Auto-ID Center faced. The MIT team also needed some world-class physics and RF expertise; for that the team enlisted Dr. Peter Cole from the University of Adelaide in Australia. With this team in place, the Auto-ID Center set about creating technology that was smaller, cheaper, and solved the interoperability issues.

The Auto-ID Center eventually morphed into a conglomerate of six universities: MIT, the University of Cambridge, the University of Adelaide, Keio University, Fudan University, and the University of St. Gallen — each with its own focus and unique capabilities but linked by the common vision of an "Internet of things." Their collective goal was to "develop new technologies and applications for revolutionizing global commerce and providing previously unrealizable consumer benefits."

In October of 2003, the Auto-ID Center licensed its intellectual property to the EPC to the UCC — the Uniform Code Council (the bar code folks). The UCC created a new subsidiary called *EPCglobal* to manage the licensing of intellectual property and to allocate EPC numbers to end users. EPCglobal also made the final determination of what the standards and protocols look like.

In late 2000, Dr. David Brock (one of the founders of the Auto-ID Center), wrote a white paper that introduced the concept of the EPC, explaining why the Universal Product Code (UPC or bar code) needed to be replaced with the EPC. (For more on the Auto-ID Center, see the nearby sidebar, "Birth of a revolution: The MIT Auto-ID Laboratory.")

In the paper, Brock cited the UPC as one of the most successful standards ever developed and pointed out that UPC coding and labeling touch a vast number of elements in the supply chain. However, as I mention earlier in this chapter, the evolution of technology has left the venerable UPC in the slow lane. The emergence of the Internet, digitization of information, ubiquity and low cost of computing power, and globalization of business necessitates a new and better solution based on a network infrastructure that has been built worldwide for the past 20 years.

The electronic product code, which uniquely identifies objects and facilitates tracking throughout the product life cycle, was created to take advantage of widespread broadband capability, faster computing power, and cheaper data storage. The EPC was designed by the team at the MIT Auto-ID Center to be a simple and extensible code for efficient referencing to networked information: a worldwide license plate for every object ever made — an Internet of things was the vision. This went well beyond what the UPC could ever dream of doing with its limited data scheme and ability to be programmed with information only one time.

How EPC is different from UPC

A UPC is limited because it contains only the manufacturer and product codes. Figure 2-2 shows the representation of data stored on a UPC.

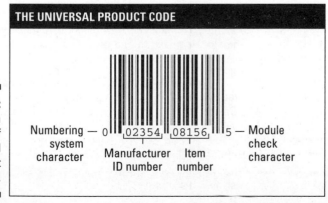

Figure 2-2:
The data structure of a Universal Product Code (UPC).

THE UNIVERSAL PRODUCT CODE

Numbering — 0 02354 08156 5 — Module
system   check
character   Manufacturer Item   character
            ID number   number

The black lines and empty spaces of the bar code hold different representation of the data than what is the actual UPC numeric code. Similar to an Internet IP address, the UPC code is made up of four different subsets:

- The first partition is a single digit indicating the numbering system used to interpret the remaining characters. For example, a *0* as the first number means that a regular UPC code will follow, but a *5* means that a coupon is going to be represented by the rest of the numbers in the UPC.

- The second partition is made up of five numbers that designate the manufacturer's identification number.

- The next five digits represent the item number.

- Because I started with a single digit, I might as well end with one. The last digit is added as a check used to validate the correct read from the scanning process — a single-digit insurance.

With five digits for the manufacturer and five for the product, the UPC can provide up to 10 unique numbering systems, 100,000 manufacturer identifiers, and 100,000 product types for each manufacturer.

The EPC differs from the UPC in a number of ways. First, like many new and improved versions, it's definitely bigger: namely, it stores a lot more information. Figure 2-3 shows how the EPC data is structured.

ELECTRONIC PRODUCT CODE

Figure 2-3:
The EPC
data
structure.

```
01.0000A89.00016F.000169DC0
```

Header	EPC Manager	Object Class	Serial Number
0-7 bits	8-35 bits	36-59 bits	60-95 bits

The setup of EPC data is quite similar to the UPC, but there are some critical improvements:

- **Header:** The first thing you'll notice is that there's a header. This tells the RFID reader what type of number follows. The header is designed so that the EPC tag can represent a military UID or an FDA code, instead of a complete EPC structure.

- **EPC Manager Number:** The next partition is the EPC Manager Number, which identifies the company or company entity.

- **Object class:** Next is the object class, similar to a stock-keeping unit, or SKU.

- ✔ **Serial number:** Lastly (and most importantly) is a serial number, which is the specific instance of the object class being tagged. Thus, it identifies the item with the specific tag as *that* item (not just a type of item).

Depending on the total data-carrying capacity of a tag, an EPC number can be from 32 to 256 bits long. This means that rather than the 100,000 possible numbers that the UPC can accommodate, the EPC can be used for millions of trillions of unique items.

Why an EPC RFID tag doesn't contain more information

An EPC tag contains something quite nondescript: a 96-bit unique identifier. This is a really big number that will never be repeated or allocated to anything except that tag. Here are the two primary reasons why EPC numbers contain only a unique identifier, as opposed to actual information about the product:

- ✔ **Security:** The EPC numbering system has often been compared with the license plate systems that departments/bureaus of motor vehicles use. Each car has a unique license plate, but you have to have access to the DMV (or BMV) database to find out who owns the car, where the owner lives, and other private information. The EPC is the same way. Because it points to a file in a database, that file and the information it contains can be as secure as any other data store. Being able to read a tag number doesn't matter if you don't have access to the database to read what information is related to that tag. The Generation 2.0 EPC protocol even allows parts of the 96-bit number to be hashed or scrambled for even greater security.

- ✔ **Cost:** Delivering a very low-cost RFID tag was one of the primary goals of the Auto-ID Center. In order to get the cost as low as possible, the components of each tag had to be as minimalist as possible. Memory on an integrated circuit (IC) is one component that drives up the cost. The smaller the memory requirements, the cheaper the tag. Originally, EPC tags held 64 bits, but end users and academics agreed that was too limiting.

The combination of cost and security drove the architects of the EPC protocol to set up an overall system in which a minimal amount of information (a unique number) could be embedded on each tag.

The EPC protocol and data structure is the standard most widely accepted in the U.S. supply chain because of the support of Wal-Mart, the DoD, and other big players. There are still many other systems (see Chapter 21 for info on standards and protocols for RFID) that do not conform to the EPC open standard but that are useful for certain applications.

When might you use a system other than the EPC? A significant example is the active tags used by the DoD. They are based on a proprietary technology by Savi Technology (www.savi.com; a commercial supplier of active tags) that stores all the information about the contents of each container, allowing a soldier in the field to scan a tag and read what is in the container. This is practical for the DoD because there is no way to look up a database file to match an EPC number with its information out in an operational theater. Until troops can parachute out of a plane and land in the remotest reaches of the Earth with a high-speed Internet satellite connection on their hips, it is unlikely that a license-plate scheme like the EPC will be effective in that environment.

How the EPC works

The EPC covers three primary areas of standards:

- ✔ **The air interface protocol:** The air interface protocol has been the source of much confusion, consternation, and chagrin since the Auto-ID Center supported the technology. The EPC Generation protocol was just recently finalized and has gone to production after much iteration and negotiations.

- ✔ **The data structure for numbers on the tags:** The data structure, which I explain earlier in the section, "How EPC is different from UPC," is now universally agreed upon.

    The data structure of the EPC was the source of much debate among the major players during the earlier days of RFID development. The original inventors of the EPC data structure felt that the governing body, EPCglobal, diminished a significant technical advantage by changing what the various blocks were used for. The DoD also responded that the data structure was not suitable for its needs and made EPCglobal include the DoD's structure into the EPC.

- ✔ **A network to look up tag information:** This is the least accepted and least evolved portion of the EPC standards. A globally adopted network isn't in place yet because many end users are simply adding EPC data to their existing EDI systems.

What are the different protocols?

As RFID has developed into a viable replacement for the bar code, several different protocols have become available on the market, and each new protocol generally improves upon the shortcomings of its predecessors.

Today, the companies requiring their suppliers to implement RFID have standardized on one protocol: the EPC Generation 2.0 (more on that in a moment). However, you still see tags using earlier protocols, so it's helpful to know how to recognize them. Table 2-2 outlines what the different protocols are and key information that you need to know about each one.

Table 2-2   RFID Interface Protocols

Protocol	Corresponding Frequency	Capabilities	Pros	Cons
Generation 1 Class 0	UHF	This is a read-only preprogrammed tag, which means that the end user can't write a new number to the tag.	Fast data communication protocol.	Preprogrammed tags increase administrative and logistics cost of affixing the correct tag to the correct item and also minimize a tag's flexibility.
Generation 1 Class 1	UHF and HF	Write once, read many (WORM)	Keep data in sequential order; manage data easier.	Can be written to only once.
ISO standard	LF, HF, and UHF	Read Only Tag Identifier with read, write, and lockable user memory to store object identifier and information.	Keep data in sequential order; manage data easier.	Does not account for the data structure but only how the tag and reader communicate.
Generation 2.0 Class 1	HF and UHF	WORM	Keep data in sequential order; manage data easier. More globally accepted protocol.	Can be written to only once.

The origins of the protocols

The Auto-ID Center identified the necessity for an RFID protocol that enabled an ultra-low-cost RFID tag implementation. The Auto-ID Center identified the cost of $.05 as the psychological barrier cost that would enable widespread item level RFID tagging. Although none of the existing standards or proprietary protocols at that time enabled such a low-cost tag, Alien Technology, a small California-based start-up, had an efficient, high volume, chip-to-antenna assembly technology. They accepted the Center's challenge of developing a low-cost RFID system with the target of a tag costing less than $.05.

The Auto-ID Center, in conjunction with Alien Technologies, developed the Generation 1 Class 1 standard. A similarly innovative company, Matrics (purchased in 2004 by Symbol), came forward with a less flexible but high-performance protocol, with product available approximately one year before the Class 1 product, and the Auto-ID Center classified it as a Class 0 stan-dard. Both Alien and Matrics focused on UHF solutions. A third company, Philips, in conjunction with the Auto-ID Center, developed a high-performance HF protocol, and it was also coined as an EPC Class 1 standard. Then ISO also got in the mix with its 18000 specifications. The primary problems in the market place, which were slowing adoption, were multiple published standards, and none with products available from multiple vendors.

As recently as 2004, the "acceptable protocols" for use in pilots and implementations were defined as either Class 0 or Class 1 Identity tags. Although these were called open standards that any vendor could use to identify products, they were based on individual companies' propri-etary technology. At the same time, the ISO 18000 standard began to emerge. However, these protocol standards are not interoperable. Clearly, the world of RFID needed a single solu-tion everyone could agree on and implement — that would be Generation 2.0.

As you make decisions about your RFID system, two factors that come into play are where you are going to incorporate the RFID network and with whom you'll have to interoperate. Different regions or countries have different stan-dards, and ISO and EPCglobal have different requirements incorporated into their open standards. Chapter 3 covers using RFID in different countries.

The Class 0 tag is Symbol's proprietary preprogrammed tag protocol, which disallows an end user from writing a new number to the tag. Preprogrammed tags both increase administrative and logistics costs of affixing the correct tag to the correct item and minimize a tag's flexibility, but remove the step of needing to write to a tag on the factory floor. Manufacturers are allocated specific blocks of EPC numbers (from the Uniform Code Council's governing division, EPCglobal) and add their own product codes and serial numbers to their assigned manufacturer numbers to create unique identifiers. Class 0 has gained most of its popularity for closed or tightly managed systems like air-port luggage tracking. However, because it was an early Auto-ID standard, Wal-Mart and the DoD agreed to support it for only a limited time.

Class 1 tags allow end users to write serial numbers to their tags, but the Class 1 Generation 1.0 reader technology does not enable communication with Class 0 tags. The other problem with Class 1 Generation 1.0 tags is that only 64 bits of memory are available on the tag, thus limiting potential numbering schemes.

The emerging ISO standard has four primary components, but ISO 18000-6 is the one dealing with the UHF range. The ISO standard differs from the EPC standard in that the ISO standard addresses only the air interface (how the tags and readers talk to each other), while the EPC standard addresses other components of the system beyond the air interface.

Working with the protocols: Why Generation 2.0 is better

The Auto-ID Center classified tags as Class 0 through Class 5 based on their functionality. Class 0 and Class 1 were designed to be low-cost identity tags and became the main drivers around the RFID revolution as they became the focus of the retail and DoD mandates. The issues surrounding the confusion between Class 0, Class 1, and the nuances in those classes have led the EPC community to demand a clarification. However, the solution to keep the same classifications — Class 0 through Class 4 — creates a next generation of each class, aptly named Generation 2.0, or simply Gen 2.0. Therefore, a Gen 2.0 Class 1 tag is still a write once, read many (WORM) tag; however, it has a truly open standard that any manufacturer can produce tags based on.

The next generation of EPC protocol is better for four primary reasons:

- It creates an interoperable, global standard.
- It makes deployment of many readers easier for end-users.
- It includes additional features that make it technically more advanced.
- It uses more advanced anticollision protocols for faster, more accurate performance.

The Class 1 Generation 2.0 protocol is backward-compatible for Generation 1.0 Class 1 and Class 0, and replaces the specifications for both classes of tags. Class 1 Generation 2.0 protocol also operates with the emerging ISO 18000-6 standard protocol (it is expected to become the ISO 18000-6c protocol with ratification in late 2005), creating one global standard and enabling an efficient solution for the lack of interoperability between Class 0 and Class 1 tags.

The Generation 2.0 tags utilize four distinct memory banks:

- **Object Identification (OID) – EPC Data:** The OID memory stores the identifier of the object to which the tag is affixed and consists of a 16-bit protocol-control parameter, a 16-bit cyclic redundancy check (CRC16) that ensures that no errors in data have been communicated from a tag to a reader (with an accuracy rate of 99.998 percent), and an object identifier that is an N-bit EPC code (where N is any valid EPC length).

✔ **Tag identification memory:** The tag identification memory (as well as the user memory) is incorporated within the Generation 2.0 Class 2 Higher Functionality tags, which allow tag- and vendor-specific data storage. The tag identification memory stores the unique identifier for the tag.

✔ **User memory:** The user memory allows user-specific memory storage.

✔ **Reserved memory:** This is used for system parameters, such as a password.

Generation 2.0 Identity addresses the folks who are concerned about privacy by having a feature in the form of the Conceal function and maintaining the Kill function also found in the Generation 1 Class 0 and Class 1 protocols. Concealed tags do not communicate any of their data until the reader sends a password to the tag. A 16-bit random number generator (RN16) is used to generate numbers for encryption of data communicated to the tag, so this number "hashes" part of the data like scrambling a bunch of eggs, making it nearly impossible for someone else to read.

How the EPC prepared for the future, and who oversees that

Sanjay Sarma of the MIT Auto-ID Center originally proposed a way of developing an open standard protocol, which today is known as the EPC. He based his idea on the work of the group that created the World Wide Web, aptly named the World Wide Web Consortium (W3C; www.w3c.com). Sarma developed EPC so that it allows true standard interoperability. The Auto-ID Center hinged its future — in fact, its very existence — on the notion that its founding sponsors would accept the EPC. Because many of the founding companies were using UPCs as the Auto-ID standard, it seemed like a reasonable bet that these companies would adopt EPC because it was designed to be a close cousin.

The flexibility of the EPC also ensures future use and global uniqueness indefinitely. A 96-bit number enables the theoretical identification of nearly a million, trillion, trillion objects. To put this in perspective, if that many golf balls were placed in a line, they would extend far beyond the edge of the known universe. If these golf balls were formed into a sphere, they would make an object six billion times larger than our sun. With that in mind, the industry believes that a 96-bit code will suffice for object identification well into the future.

In November of 2003, the Auto-ID Center turned over its work to EPCglobal, Inc., a joint venture between EAN International and the Uniform Code Council, Inc. According to its Web site (www.epcglobalinc.org), EPCglobal "carries forth the research completed by the Auto-ID Center to work toward the development of industry-accepted standards and commercial adoption."

With the various standards evolving and more people requiring their trading partners to adopt forms of RFID, keeping an eye on what is happening at EPCglobal can be critical to your long-term successful adoption of RFID. EPCglobal also has several highly involved working groups, which bring end users together to help figure out the best way to adopt RFID, from data standards to preparing analysis on return on investment (ROI).

Addressing Privacy Concerns

With any new and highly disruptive technology, issues are raised about the potential downside. The first among these, for the case of RFID, is the concern about individual privacy:

- **Why people are worried:** A considerable driver of the fear of RFID as an invasive technology is a lack of understanding about it. In a recent interview, I was asked about sticky RFID tags that could surreptitiously be placed on the bottom of someone's shoes and then tracked wherever they went, which is impossible. A more realistic fear is that someone could monitor the arrival and departure of an item you've bought. For example, if you purchase a watch with an embedded RFID tag and the tag was associated with you, someone could conceivably monitor every time you came and went to that store. However, this collection of information is similar to what happens today. For example, if you pay with a credit card or if you use a grocery discount card, that store is gathering information about all your purchases. The concern is an extension of an old issue.

- **Why people shouldn't be concerned:** The simple solution in most instances is the basic countermeasure employed by everyone but Minnie Pearl — that is, just take off the tag. If the tag is actually embedded in something, cutting it off or just scraping the chip off the tag (easily done with a pair of nail clippers) is sufficient to ruin the tag's ability to communicate. The infrastructure and cost of RFID networks are also prohibitive factors. Simple return on investment dictates that no amount of information will pay for investing in an infrastructure that would be ubiquitous enough to track items everywhere you go.

Although the U.S. Federal Trade Commission (FTC) and Congress have begun instituting hearings to consider the possible regulation of RFID frequency bands — which currently use unlicensed bands — existing property rights and common-law privacy torts substantially limit the potential abuse of RFID. Putting a chip or a reader on a person or person's property against his or her will is a tort and a violation of constitutional rights and fundamental liberties, which is certainly a criminal offense. Hiding RFID readers in your house or car would be no different than wiretapping your phone or hiding a video camera in your kitchen. (And an RFID reader is a lot easier to locate than a wiretap because it operates at a known, easily detectable frequency.)

Social forces will most likely regulate the use and deployment of RFID long before the government and lawmakers get involved. The various types of forces that regulate emerging technologies include economic incentives (using a speedpass RFID token to shorten your wait at the gas pump), consumer preferences (not having to stand in line to upgrade your DVD player with new software), countermeasures (new tag protocols that cause the tag to self-destruct), and existing laws that regulate the use of personal information. RFID technology will be no more and no less controlled by these forces. Among these regulating forces, buyer preference will most effectively regulate the use of RFID. Some citizens won't want to pay to create and enforce laws and regulations; others may opt for the convenience RFID offers.

Activists who have considerable privacy concerns should focus on ensuring that companies post a simple notification that they use RFID tags, educating consumers of the true technological limitations, and tearing up all their credit cards (because that's where the juiciest information can be tracked, like everything a person purchases online).

Chapter 3

Making Basic Decisions about Your RFID System

*D*eploying an RFID network is a little bit like farting in church: It has specific impact for one person, it affects everyone else around that person, but not everyone is going to stand up and say how they feel about it. The difference is that, if you plan things correctly, your RFID deployment can be a big hit with everyone — if you figure out how to make that happen with a fart, let me know.

To make your RFID deployment successful, you need to consider all the stakeholders and understand the different areas impacted by a new RFID network before you start putting systems up around your facility. This chapter walks you through the basic decisions and assessments you need to make your deployment a success.

✔ First, I explain how to examine the areas of impact within your organization.

✔ Then, I cover some basics of the technology, so you can begin making basic decisions that shape your RFID network. I talk briefly about the frequencies of various systems, and then give you some performance criteria to think about as you plan your RFID strategy.

With just a little understanding of the areas impacted by RFID and the types of systems available, you can start to understand the performance constraints of modern RFID systems. This knowledge can also help you determine who to involve in your RFID project and, more importantly, this knowledge can help you avoid many of the "gotchas" that early adopters faced. If you begin your design with the end in mind and you roll in the laws of physics, the limitations of the current technology, and how products behave in an interrogation zones, you've got the key ingredients of cooking up a successful RFID network.

Midas Touch Points: Where RFID Impacts Your Organization

Without a good understanding of certain components within your business, you can't take advantage of RFID's ability to provide real-time, serialized data. You merely get more bad information a lot quicker — no more use to you than a bad smell in a crowded room. You can have the Midas touch by turning this new technology into gold for your organization if you understand the basics of the business process change, the impact on the information systems, and what needs to change within each facility.

The best way to get your arms around how the RFID network will impact your organization is to assess your business from the touch points RFID creates. Knowing those points is your key to success (and a critical part of the four Ps — Process — which I introduce in Chapter 1). If the planning is done correctly, these touch points can be as good as gold for your organization: They'll give you information about your systems, processes, and enterprise, creating a true competitive advantage.

For most businesses, you can break this assessment of RFID's overall impact into three distinct workflows: business processes, physical infrastructure, and systems and technology. To map out the points where RFID has an influence (the *touch points*), here's an overview of the steps to follow:

1. **Map out your current business processes and then identify points on that map where RFID can improve those processes.**

2. **Determine points where you need to change your physical infrastructure in order to accommodate RFID.**

3. **Examine your current systems and technology and identify points where you need to add to or change your IT infrastructure to accommodate RFID or fully take advantage of its benefits.**

Discovering the touch points for these three workflows is your key to involving the right people, picking the best starting point, creating a successful long-term plan, and buying the right hardware and software.

This exercise alone can help shed light on what's happening in the organization and is a good way to get the RFID team working together. In Chapter 15, I cover how to organize an RFID team, which is of critical importance as you go through this mapping to make sure you're not overlooking any areas of your business.

Each area (business processes, physical infrastructure, and IT infrastructure) must also be linked to your overall strategy. That is, you must also ask the question, "What does the end-game look like?" As you work through each area, you should gain insight to the following kinds of strategic level questions:

- After RFID is fully deployed, what is my competitive edge?
- How does this RFID deployment align with my supply chain management goals?
- Do my trading partners share similar goals and objectives regarding RFID deployment?

The following sections explain each step of the process in detail.

Outlining how RFID affects your business processes

No matter what type of RFID application you're using, you can map it as a business process flow. Make sure you evaluate your current business processes from end to end. The easiest way to start this is to get a whiteboard (or an easel with sheets of paper) and diagram what happens from the earliest point in your production or supply chain as possible. Here are some common examples:

- **Raw goods:** Begin mapping from the point raw goods hits the arrival docks. Show how the raw goods process changes into work-in-process, and then what happens as it leaves the facility.
- **Distribution channels:** Map what happens when products arrive, how are they stored and sorted, and what happens as they leave.
- **Asset management:** Begin when something is ordered and map what happens when it arrives and how it's accounted for annually.

The biggest ROI (return on investment) and strategic benefits from RFID come from being able to improve your business processes. Trying to deploy RFID with no change in existing process means only that you're adding an expense. That's why, after you map current business processes, you need to identify the business process touch points — that is, the points where RFID will improve your business process. To identify possible touch points, look for places where

- **You can automate human tasks.** Any place where human intervention is required to read a label or scan a bar code is a possible place to put an antenna and RFID reader to automate the procedure. By doing so, you can eliminate employees and

✔ **You can eliminate frequent data errors.** Data errors are usually the result of humans doing the counting, so look at all the human interaction throughout the process.

✔ **You can benefit from real-time data capture.** For instance, if there's a critical point in the working process where things need to be ordered or work has to be scheduled, knowing when something reaches that point is critical. RFID can enable that knowledge by sending specific information or instruction to a central application the moment it reads a tag.

✔ **Things must pass by a *choke point* (a place where all items need to pass through during the normal process, like a toll booth on a highway is a choke point for traffic).** Reading data at these points will give you an idea of real-time volume flow and timing. If you have applications that are set up with automated decision rules (such as, do something if a certain condition exists), RFID can enable these actions without human intervention and in real time. A good example is scheduling a warehouse pickup; if you are getting RFID information from a location at certain dock door, you know that the next truck coming into the warehouse needs to be scheduled at a different dock door.

✔ **Things are in-bound from suppliers or partners.** Eventually your partners will be tagging things, and if you can automatically count those things, you can streamline the put-away, billing, and forecasting processes significantly.

✔ **You can directly correlate with key interrogation zones (places where RFID readers scan tags for information) to check accuracy of information.** In other words, if things are put on a specific conveyor after they come in a dock door, spending a couple thousand dollars for another interrogation zone and linking it to the initial interrogation zone (at the dock door for instance) can insure a high degree of accuracy and let you know when one of the readers has performance issues.

For analysis purposes, try to incorporate RFID everywhere, even if you know that it doesn't make sense from a financial or technical perspective. You may find possibilities you haven't thought of in terms of benefits to the business process. Brainstorm the RFID possibilities — don't try to justify them or consider the feasibility just yet. Think about where you might benefit from automatic data capture and automatic counting.

As you work on evaluating business process touch points, the following questions may help you translate all the information you gather into a more concrete analysis:

✔ What objectives will determine success for your RFID initiative in the next 6, 12, or 36 months?

✔ What benefits can RFID deliver and what cost is required?

- ✔ How will you measure results against your baseline?
- ✔ At what facilities and which products will you pilot first?
- ✔ How does the business process need to change to accommodate RFID?
- ✔ What user scenarios and processes will you pilot?
- ✔ How will you document the impact that the pilot has on business processes?
- ✔ What personnel issues need to be addressed — safety concerns, union questions, privacy, and education?
- ✔ Should the program be funded with capital budget or operating budget?
- ✔ What are the tax implications of an RFID network?
- ✔ Can you get your landlord to subsidize it as part of the build-out allowance?

CASE STUDY

The case of the missing pallet

One large U.S. retailer has wrestled with inventory tracking issues concerning outbound trailers and full pallets of items. The existing system for flat-loaded cases (those that employees put directly on the truck, one at a time) *should* work something like this: An employee scans a case bar code as it is loaded into the truck from a conveyor going into the back of the truck. This bar code scan as the case actually goes into the truck indicates that the case has been loaded on a specific outbound truck. In reality, though, pallets are not scanned as they go onto the truck. Instead, the pallets are staged off to a dedicated area in the warehouse *near* the outbound trailers — but not directly in them — just waiting for the truck to be almost full so the pallets can be loaded last, and therefore unloaded first. When the pallet is staged off in that special area, the employee scans a bar code that associates the pallet with the specific outbound trailer.

The problem this retailer faced is that often the pallet doesn't make it onto the truck because the truck fills up with individual cases being flat-

loaded. In the inventory management system, however, the pallet is recorded as having been shipped on that truck and then received at the store where the truck is off-loaded. This creates a significant cascading effect on inventory accuracy and data collection.

A situation like this is the perfect opportunity for RFID to automate a bad process and improve inventory accuracy and data collection. This problem begs for RFID readers at the dock door, logging everything that gets put onto the truck — and only things that get put onto the truck — by scanning dedicated pallet tags as they pass by the readers. Even though the pallets may be temporarily placed someplace for later loading into the truck, they will be read only if they make it on the truck. Not only will RFID make the process much more accurate, it will eliminate the business process of the person dropping off the pallet having to stop and scan a spot on the wall with the truck bar code to make the association between truck and pallet. The forklift driver can simply drop off the pallet at the staging area and be on his merry way.

Determining how RFID will affect your facility

Looking at your business process flows (as described in the preceding section) spawns creative places to put RFID *interrogation zones* (radio frequency fields where the tags are to be read). The next logical step is assessing touch points in your physical infrastructure — or how and where your physical infrastructure needs to change in order to accommodate those interrogation zones.

As you look for these touch points, consider the following questions:

- **What new hardware will you need?** RFID systems require antennas, readers, routers, data cable, and so on.

- **How can you install, ruggedize, and power those systems?** Look for infrastructure that's already in place (power and networking) as well as capabilities you might need to add.

- **How will data be sent back to a central application if you actually set up readers and antennas in those locations?**

- **What items will you be tagging and tracking, and how and where will you tag items?**

- **What performance and tracking requirements must be accommodated?**

- **What other systems in place currently generate RF noise, and how can you plan to reduce that noise?** In addition to your own facility, consider how your systems might affect your neighbor's or how your neighbor's systems might affect yours. Consider how you might make expansion facilities RF-friendly.

You can find more information about determining what hardware you need and how to set it up in Chapters 9 and 10.

The assessment of how RFID will impact your environment is the perfect opportunity to involve your facilities team — engineers, property managers, electricians. These folks can help you determine where you might need custom installation solutions and understand whether you need to add things such as additional conveyors, structures to protect against forklift accidents, and additional power sources.

A great way to begin understanding the impact on the physical infrastructure and facilities is by practicing a little *management by wandering around* (MBWA). After walking through your facilities with your warehouse or production supervisor, get a copy of your engineering drawings and mark out potential locations for the RFID interrogation zones. This helps you understand where you may have to supply dedicated power, install CAT-5 cable, and alter current working conditions like conveyor directions or forklift paths.

In addition, the information you gather in MWBA helps when you make technology assessment decisions, as described in the next section. For example, you may decide that you need a reader that can be powered over Ethernet, or that RS-485 is the only communication that will make it across your full facility.

Evaluating your technical needs

The four little letters, R-F-I-D, have shot fear into the heart of system administrators and application engineers like nothing since the three characters, Y-2-K. Many early estimates of the impact of RFID technology projected that terabytes of data would surge out of each warehouse, and companies would have to install dozens of T-1s to transport the data between locations. Fortunately, that isn't the case for well-architected and -constructed RFID networks.

A critical step in creating this well-constructed network is understanding all the touch points of an RFID network to the IT system. Because the benefits of RFID networks are predicated on transporting real-time, serialized data into an application that can make sense of it all, two areas to examine for IT touch points are applications and data transmission.

Applications

One of the keys to handling the billions of reads that happen in a typical warehouse is to have specialty *middleware* (filtering software) for the readers. The data created by an RFID reader needs to be filtered and smoothed before it is useful for any application. The special filtering and smoothing software is called a *reader interface, EPC information server,* or a *Savant.* The reader interface is the key to making the data volume manageable and the information useful. (I go over the various middleware options in detail in Chapter 11.)

The middleware enables you to use RFID with existing applications, such as warehouse management systems (WMS), enterprise resource planning (ERP) suites, or inventory management packages. The good news is that the leading software companies are all developing middleware modules specifically designed for RFID. Although many of these will not be commercially viable until 2006, they will still solve many of the problems IT staff are concerned with, such as the form that the data will need to be exported in, how the multiple reads will be filtered, and what the key filters for the data are.

Data transmission

Data transmission is another critical area to consider beyond the principal business applications. You need to figure out how to transmit the data coming off your readers to your trading partners — most likely on one of your existing systems. Although this is nothing new, you do need to transfer more data and do so in a different format from the UPC. The key RFID influencers who are driving the mandates — Wal-Mart, the Department of Defense, Tesco, Target, and others — require partners to

- ✔ **Use the electronic product code** (more commonly known as EPC; see Chapter 2 for details).

- ✔ **Exchange that EPC data by using existing systems** such as EDI (Electronic Data Interchange, the systems used to send order information between partners), Web-based systems like AS2, or custom systems like Wal-Mart's Retail Link.

This data exchange happens today without EPC numbers: When a client orders a particular item from a supplier, that supplier sends an order confirmation and/or an advanced shipping notification (ASN) back to the client. The client company stores that information — specifically the item's description and UPC or serial number — in its system locally, and knows to expect the item in an upcoming shipment.

To optimize your systems for RFID, the critical task for your organization is associating the scanned EPC numbers as they come in the door with EPC numbers on things like an ASN. Part of determining your IT touch points is figuring out how to use your existing methods of data interchange to carry additional information about each item, case, or pallet received. In most instances, this is a simple 96-bit hexadecimal number (the EPC number), so you should plan to deal with that number as a unique data identifier.

Some people have talked about a global look-up system like the Domain Name Service (DNS), particularly VeriSign, which runs the DNS system. VeriSign has invested heavily in building and marketing an EPC network specifically to look up EPC data. However, this is likely to be used only by a few niche industries, such as the pharmaceutical industry. Most other industries will have existing methods of transporting data and won't go to the additional expense of adding another network layer that requires each number to be looked up. It is seldom necessary to look up each EPC number on a central data repository like you do with a Web page or other system using the Domain Name Service (DNS). Data will travel the way it does today — by being *pushed* through existing systems, which will be optimized for RFID. A look-up system like the EPC network would only be used for exceptions. The other solutions that exist are specialized overlay networks for global data exchange. By only looking up exceptions and pushing EPC data over existing systems like AS2 or ASN, the bandwidth requirements are significantly lower than if you looked up every tag that was read coming into your facility.

Drilling down to the touch points

With an overall understanding of how your IT infrastructure needs to change to take advantage of RFID, your RFID team can begin assessing more specific

IT touch points. The following questions can help you make critical decisions and create a well-developed implementation plan:

- **How will data in remote locations away from primary data centers be stored and backed-up?** Estimate the data storage needs for the next 6, 12, and 36 months.

- **How will data be transported between locations?** More importantly, what data has to make it back to a central application or location to be actionable, and what form does that data need to be in? Consider whom you need to synchronize data with, and how you'll address data synchronization internally and externally.

- **What is the appropriate integration strategy and technology to employ?** Consider how RFID will impact the existing applications. Do you need to do a full integration with your and your trading partner's applications or can you take advantage of new technologies like integration networks from providers like Grand Central?

- **How will you manage the RFID infrastructure after it's installed and operational?** Do you need a 24/7 operational RFID network? Consider whether you need to increase bandwidth to facilities with RFID infrastructure and whether you need to rearchitect your network to include wireless or additional security for RFID?

- **How will configurations be classified for different reader set-ups?** How can those configurations be pushed out automatically to new readers added to the network? You'll want untrained personnel to be able to add readers or replace faulty ones, and know that they belong to a configuration class that may be called dock door, or conveyor, and then push that configuration remotely onto that reader.

- **What happens to reader configuration, data, and routing if a power failure occurs — is there a back-up strategy in place?**

- **Can the RFID system be incorporated with logistic systems to allow trucks inbound with tagged items to specific dock doors that are set up with RFID?**

- **How will existing processes need to be changed to handle data?** Does information come back today in batch form once a day, once a week, or at other scheduled intervals? RFID's benefit is giving information back in real time — what changes need to be made to leverage that advantage?

- **How will your systems take advantage of serialized data?** Most bar code scanners can record only that a box of tiddlywinks arrived. It won't likely know the difference and specifics of each individual box of tiddlywinks. A significant benefit of RFID is that you can know a vast amount of data on each box, case, or item, such as when it arrived, where its going, how long its shelf life is, and so on. Being able to act on this type of data is critical to your IT system.

Data synchronization

A core component of the success of RFID networks in a global, collaborative world is the ability to synchronize data with partners. That data may be anything from case docket numbers for evidence to SKU numbers and pricing. Various industries are looking at different ways of synchronizing the data between trading partners, but it is a challenging issue. The pharmaceutical world, for example, could have one central database to list products and track the location of each of those products, but who would manage that? The Food and Drug Administration is a likely candidate because there are significant security, fraud, diversion, and liability concerns that RFID can solve without giving away competitive information; however, the FDA is not set up to do that. Moreover, they don't want to do it.

The logical place to turn is the associations. The Healthcare Distribution and Manufacturers Association (HDMA) is one possibility. The goal is to have a central repository where stores, distributors, and hospitals can verify the pedigree and chain of custody of high-theft or high-counterfeit drugs. This is why a central database would need to be accessed by many different parties. This issue is prevalent across many industries starting to look at RFID and exploring a more collaborative network among trading partners.

The retail world is leading the charge and may be a good place for others to look for early learning. The retail industry has standardized on the EPC protocol and data structure on each tag so there can be a central store of data: the EPCglobal registry. The registry exists so that accurately described and consistent product information can be exchanged between trading partners. Wal-Mart has selected the Uniform Code Council and its data synchronization system UCCnet as the preferred method of exchanging information. As of this book going to print, 650 suppliers send item information to the data pool on a machine-to-machine basis. This model seems to be working and will eventually be replicated across other industries.

What's the Frequency, Kenneth?

REM didn't know the first thing about RFID when they sang that line in their '90s hit, but it's the right question to ask (if you ever find out who Kenneth is) because the choices are many.

The mandates many suppliers are forced to comply with will dictate the type of RFID network you put in. Most of the highly publicized mandates require suppliers to use UHF, which is 902–928 MHz in the United States.

The reason for this choice of frequency is largely due to work at MIT's Auto-ID Center in early 2000: The innovation centered on UHF's ability to read tags at greater distances than other frequencies. This has made UHF the early favorite in the supply chain world. However, other frequencies have their places in the evolving world of RFID.

Understanding the difference between licensed and unlicensed frequencies

Because RF waves can have such a big impact on other receivers and are much more difficult to block than say, light waves, RF waves are tightly controlled by the FCC (and its equivalents in other countries). The FCC and its spectrum license process (which you may have heard about in spectrum auctions) specify the frequencies, communication means, amplitudes, and uses that are permitted through a process known as *spectrum licensing*, which allows for both unlicensed and licensed bands:

- **Unlicensed frequency bands:** The popularity of particular frequencies arises because of pre-existing technology and the fact that the Federal Communications Commission (FCC) has allocated certain sections of the frequency band as *unlicensed*. This means that as long as people follow rules of transmission and broadcast, anyone can use these frequencies. The FCC has created unlicensed usage for low-frequency (LF), high-frequency (HF), and ultrahigh-frequency (UHF) bands.

- **Licensed frequency bands:** These bands require you to pay to use a certain frequency band. Ever since governing bodies began allocating licensed frequency ranges, big-spending speculators have been cashing in. Frequency bands were the area of much speculation during the '80s and '90s and resulted in many spectacular overnight millionaires.

The good news for those folks who need to adopt RFID because of a trading partner or client requirement is that they will likely be able to use the unlicensed ISM band and not have to deal with the FCC. Chances are you need to use UHF in the unlicensed ISM band (902–928 MHz), and you won't need to worry about purchasing spectrum rights. If you're looking to build proprietary closed-loop systems, you need to go through the process of applying for an FCC license. Using a licensed spectrum is applicable for closed systems in which interoperability is not of concern.

Examining the most common frequencies in RFID

RFID systems use many different frequencies, but generally the most common are unlicensed low-frequency (LF, around 125 KHz), high-frequency (HF, 13.56 MHz), and ultrahigh-frequency (UHF, 902–928 MHz). Microwave (2.45 GHz) is also used in some applications.

Radio waves behave differently at different frequencies, so you have to choose the right frequency for the right application. Here are two key factors to keep in mind:

- **How far the frequency is able to read:** The higher the frequency, the shorter the wavelength for RF transmission. Without going into the details of the physics, the shorter the wavelength, the better a small antenna like an RFID tag is able to receive a transmission at greater distances. Therefore, the lower the frequency, the shorter the read distance for an equal tag size.

- **How well that frequency works with the materials you need to use:** You need to decide the properties that are most important in the application, such as being able to work in an environment where metal, liquid, or machines are prevalent. See Chapters 9 and 10 for more details on how different materials affect RF waves and how to get optimal reads with those materials.

Table 3-1 gives an overview of the read distances for the different frequencies as well as examples of the types of applications that might use particular frequencies.

Table 3-1	Frequency Applications for Passive RFID	
Type	*Read Distance*	*Applications*
LF (low frequency)	Reads at very close range, just beyond actual contact.	Access control and payment technologies. Usually not used for tagging objects.
HF (high frequency)	Can move out to several inches — around a foot with good planning.	Many items in close proximity, like pharmaceuticals on a shelf. HF works well on liquid medicine vials and similar products.
UHF (ultrahigh frequency)	Can easily read several yards and, in a perfect environment, 30 or 40 feet and beyond. Because a typical dock door is ten or twelve feet across, UHF is the darling of the Auto-ID world.	Supply chain, asset management, and access control for vehicles. UHF has challenges with direct contact on liquids and metals because the frequency is easily reflected and absorbed.

Frequencies, power, and countries

UHF RFID systems have been around only since the mid-1990s, and countries have not agreed on a single area of the UHF spectrum for RFID. If you're part of a global organization that's planning on adopting RFID worldwide, you need to look at the various frequencies and power levels being used across the globe. In North America, Europe, and Japan:

- ✔ **Low Frequency (LF) 125–134 kHz** is used for RFID applications.
- ✔ **High Frequency (HF) 13.56 MHz** is used at very similar power levels.

UHF frequencies aren't so streamlined, however. Table 3-2 outlines the UHF spectrums used in different geographic areas at the time this book went to press and the power antennas require to use those spectrums (expressed in both watts and ERP).

Table 3-2	UHF Frequency Bands around the Globe		
Area	*UHF Frequency Band*	*Power*	*Notes to Keep in Mind*
United States	902–928 MHz	4 watts, 4W ERP	Unlicensed spectrum used in the United States deploying spread spectrum transmission at up to 4W ERP. This band must be shared with other (non-RFID) users observing the same frequency hopping parameters.
Australia	918–926 MHz	1 watt, 1W ERP	Australian allocation available for RFID, up to 1W EIRP.
Europe	865.6–867.6 MHz	2 watts, 2W ERP	There is a draft recommendation for extension to the unlicensed 869.4–869.65 MHz European band that would allow transmission at up to 2W ERP, dividing the spectrum up into ten 200 kHz channels.
Japan	952–954 MHz		This allocation opens for RFID in Japan in April 2005.

Also, on the other side of the Earth is the hugely influential factor in RFID frequency standards I call the China Syndrome. Nearly 20 percent of what is shipped into the United States from China is destined for Wal-Mart. Ideally all those cases coming in would have a UHF tag affixed to them, per Wal-Mart's mandate. However in China, 915 MHz is used for GSM mobile communication systems: A factory, warehouse, hotel, or community may set up its own privately managed system at this band. Because this band is already actively used in China, it is unlikely to be reallocated for RFID.

With this heterogeneous mix of global frequencies, both readers and tags have to become more agile. Now both readers and tags that can read multiple frequencies and have various power levels are on the market. However, no regulatory body (like the FCC) wants to allow a unit whose power can be changed to be sold, because enforcement issues would be a nightmare. The U.S. DoD is having to address these problems in a particularly creative way because it sets up and builds systems overseas and needs to comply with the regulations in the resident country. It's a tricky situation for any global multinational corporation or agency like the DoD to maintain central IT infrastructure control and local compliance.

One thing you may end up doing, if you're planning and testing in the United States, is applying to the FCC for a specific license to be allowed to test European or Japanese spectrum in the United States This helps you decide on the right hardware and solutions for your global infrastructure but enables you to test them in one location.

For more information, the government, as always, is here to help. Check out the FCC's site:

```
http://gullfoss2.fcc.gov/prod/oet/cf/els/help/STA_Help.html
```

Beyond UHF: Looking toward the future

UHF has undoubtedly taken the early lead in the RFID race. Investment and innovation have been centered on UHF, but the buzz around RFID is creating an investment tide that will raise all boats. Recently, deals focused on HF and 2.45 GHz (microwave) have also received funding from venture capitalists. This means that other frequencies will have an opportunity to show their strengths as they become more user-friendly.

Certain niche industries will drive the adoption of frequencies other than UHF. This is happening today in the pharmaceutical world. The big pharmaceutical companies, the Food and Drug Administration (FDA), and some drug stores are currently deploying 13.56 MHz (HF) systems for tracking expensive drugs. The reason they have decided to use HF and buck the trend of using UHF is twofold:

✔ Most of the drugs have a high liquid content, and HF performs better than UHF for close-range liquid items.

✔ A number of items in very close proximity to one another need to be accurately read. HF has a better ability to manage the issues raised by this close proximity in what is referred to as the *near field,* particularly if there is little movement.

The near field and near-field communications (NFC) are getting their own notoriety as a unique technology. However, they're really just a specific type of RFID that uses the HF band of the spectrum (in the United States and Europe, this is most often 13.56 MHz). In addition to the different band, it uses a different form of communication that has an impact on the usable range. Remember that the lower the frequency, the longer the wavelength. If 13.56 MHz is used, the wavelength is about as long as a football field; anything closer than that full wavelength distance is said to be in the *near field.* Practically speaking, this means that NFC must happen in very close range — usually under one foot. Near-field communication is all via magnetic waves and has an inverse sixth power ($1/r^6$) relationship with range, which explains why the range is only a few inches because the power to wake up and communicate with a tag diminishes very quickly.

Speed, Accuracy, or Distance — Pick Two

It is quite possible to optimize various RFID frequencies to work at extended distances, and on very fast moving items, and with many items in the interrogation zone — but not optimally with all three at once. The limitations surrounding RFID systems and various frequency choices are such that you need to decide the most important criteria for the information you need and design your system around that criteria. Another way of saying this is that you need to account for the laws of physics (one of the four Ps that I introduce in Chapter 1).

The laws of physics clearly dictate the behavior of radio frequency, electromagnetic radiation, and radio communication. Although many people would have you believe it is some esoteric black art, it is not. Many people make the mistake of not even considering some of the known laws of physics as they start building an RFID system. The systems they deploy are usually deemed poor performers because of so-called immature technology, when in fact it's because of engineering errors. Another reason people say RFID technology is immature is that they try to buy a "slap and ship" solution, thinking it will work for any application. But there is no silver bullet, or "RFID in a Box" solution, no matter what someone's marketing literature tells you.

A properly designed RFID network is customized for the particular environment, product characteristics, IT systems, and business processes. The only way you can select the right tags is to test what will be tagged. The only way to select the right readers is to investigate the physical environment, know the tags you're using, and then decide on the best solution. Although this may seem like an impossible task to manage, you can build a system that's both customized and meets your needs by designing the nodes of your RFID network one at a time, but doing so with the full network in mind and going for as much standardization as possible. Follow a sound scientific methodology starting with the physics, and you'll have a highly accurate, easily maintained RFID network.

The follow sections help you gain a basic understanding of the physics and limitations you need to work with as you design the system and how that system will work within your business. This section can be your foundation, and Part II explores the physics of RFID in even more detail.

Designing for the right read distance

Distance is one simple parameter to consider as you evaluate your system requirements. By understanding your business process, where you need to have interrogation zones, and how wide an area you need to cover (all explained earlier in this chapter), you can figure out how far your readers need to read.

If the area is wider than a couple feet, you likely need UHF (as opposed to HF or LF). When designing an UHF system, you want to determine the maximum read distance as well as the minimum because in RFID networks, too much distance isn't a good thing. Reader manufacturers make bold claims about the distance at which their readers can read a tag: "Tags read at 15 meters" or "30-foot accurate reads." These distances don't matter in the real world. In all the deployments I've done, I have yet to see a dock door that's 60 feet across (needing a reader that can read 30 feet from one side and 30 feet from the other side). In fact, very few applications require a read distance beyond 6 or 7 feet, particularly in a typical dock door setup with an antenna on each side of a door (so a read range needs to be only half the door width).

After you know what distance you need, you also need to consider the quality and design of the equipment and other factors that affect the read range. The primary factors are

- **The sensitivity of the radio on the reader or its receiving capabilities:** You can determine this by comparing readers side by side in a lab setting, which I explain in Chapter 10.
- **The transmitted wave's absolute power:** This is usually regulated by governing bodies like the FCC or ETSI.

A number of sophisticated (and well-known) physics equations dictate the effect of distance and the power of the wave transmitted on read range. An RFID wave propagating from an antenna is like a grenade exploding — it blows up in all directions. One of the known equations of physics proves that when the distance from the transmitter doubles, the strength of the original RFID signal (or wave) is quartered. The fancy name for this is the *inverse-square relationship* and knowing this relationship helps you plan what level of power you need for the distance you need to cover. I go over this in more detail in Chapter 10.

✔ **The path of the wave relative to environmental factors.** Ambient electromagnetic noise or interfering factors can affect a reader's ability to obtain accurate readers across a given distance. In Chapter 7, I spend a lot of time helping you understand how to look at the environment and all the noise that may affect your RFID system.

RFID is just another form of radio communication — like your car's radio or mobile phone — and it is susceptible to interference. That interference can come from many different types of materials. As RF waves slide through materials, they may be absorbed or reflected, depending on the properties of the material and the type of radiation. This change in the RF waves results in a reduction, or *attenuation,* of the strength of the wave. Attenuation changes depending on the type of material RF waves pass through. In addition to attenuation, there can be interference by other waveforms that desensitize the receiving antennas of the RFID system.

Bottom line is, to read tags accurately across a given distance, you need to carefully consider the readers' radio and power and the environmental factors. However, the farther away you need to read, the more likely you are to run into diminished power and interference from environmental factors.

Reads — tell me how fast and how many

After you understand the principles behind read distance and can identify the maximum read distance you need (as described in the preceding section), the next step is to figure out the speed of reads required in your system and how many can be read in a set period of time.

If you're using a bar code system today, even adding two reads per second can be twice as good, or 100 percent better, than what you have today. RFID allows much more than double the bar code speed.

To figure out required speed for your tag readers, you need to understand how many items will go through the interrogation zone per second or per minute. Basically, what you're trying to determine is whether the solution needs to be set to interrogate continuously so it never misses an item (for example if you need to read pallets as they pass through a dock door), or if you can just look for changes in the field (like if something was sitting on a shelf or in a case). If you're not sure, see Chapter 6, where you can find more details about common RFID setups, such as dock doors, conveyors, shelves, and shrink wrap stations.

If you determine that you need to interrogate continuously, note that an important factor in the speed is the air interface protocol that your system uses. If your system is tuned and configured properly, you can read tens of dozens of RFID tags in an interrogation zone *almost* at the same time. It happens so quickly that it may seem like the tags are being read simultaneously, but the majority of air interface protocols (like the modified ALOHA slot protocol in the Generation 2.0 standard) are read and identified consecutively so that there are no issues with *collision* — tags interfering with each other.

The anticollision protocols in the Generation 2.0 version of the electronic product code (EPC) standard were under much debate in order to attempt to optimize the reads per second. The critical factor for performance is that the protocol and readers can identify unique tags. The reader determines this by *talking first* — or sending out a signal to the tag — to wake up the tags in a method known as *reader talks first*.

I introduce the EPC and its protocols, including Generation 2.0, in Chapter 2. For more details about how these protocols work, see Chapter 5.

Reading multiple tags at once — accuracy considerations

Like in most things in life, faster is not necessarily more accurate in an RFID system. As you evaluate tag readers, you want a metric that accounts for the speed and the accuracy.

Thus, the best metric for comparing different readers is the *time to last tag in the field*. The system that best "wakes up" *all* the tags that you need to read in the interrogation zone (as a pallet comes through a forklift or a tray leaves a centrifuge) and transmits the information on those tags back to the receiver is the system that is more accurate and will best meet your needs.

Make sure you distinguish the "time to last tag in the field" metric from the *absolute best value* for the numbers of reads per second. For example, it's possible to read a *single* tag several hundred times in a second (the *absolute best value*), but that same system may not be able to read a hundred *different* tags at once in that same second.

You discover the difference between the two metrics in your reader testing, which I explain in more detail in Chapter 10. Here's an overview of the steps to clarify how the process works:

1. **Test a single tag in the interrogation zone and record the number of reads over a set period of time.**

2. **Test some number of tags over the same period of time.**

3. **Compare the two read rates.**

    You can expect results something like this:

    - 1 tag tested for 30 seconds yields 1,000 reads

    - 10 tags tested for 30 seconds yields 100 reads

In addition to the metrics, here are some considerations for weighing accuracy against speed and distance, as you decide on the basic design of your RFID system:

- If tags move through an interrogation zone very quickly, you need readers configured to broadcast and receive data very quickly. Because that setup is focused on very quickly reading every tag that goes by, it is not possible to read more than a few tags at maximum speed.

- Similarly, because a tag takes time to respond to an inquiry from a reader, the farther away that tag is from the reader and the weaker that signal is, the longer it takes to get that response back and receive the next one.

So if you want speed and accuracy, you need to keep the number of tags to read within the capabilities of the system, and if you want to emphasize speed and accuracy, you want the shortest distance possible between the tag and reader. The reason for these constraints is because of the *frequency hopping* — the fact that each channel can broadcast only for a short period of time and then has to switch channels. Therefore, if you're in the middle of communication and the channels change, there is a higher likelihood of missing a receiving communication from the tags.

Now What about the Tags and Objects?

The tag is in fact a tiny RFID system: It receives an RF wave, processes the signal, and transmits an RF wave. Those waves are subject to the same laws of physics that the readers are subject to. With these laws in mind, you need to consider how the objects and the tags on those objects will work with your system:

✔ **Consider how the object might affect the system:** Whether the object is a case, a pallet, or a laptop, you need to know how it will respond to the various factors that make up an RFID system. If the tag is surrounded by liquid, for example, that liquid will strongly attenuate the RF wave. A metal object may change the *tuning* of a tag (or frequency on which it can receive signals), reflecting the RF waves from a reader, or block communication from a specific antenna.

✔ **Remember that the tag and object need to work together while frequency hopping:** The constraint that you have to face in the United States is the use of *frequency hopping* across 124 channels from 902–928 MHz. This means that the tag on an object has to be properly tuned to read well in all those channels because, according to FCC regulations, a reader can broadcast on any given channel for only a certain period of time (400 milliseconds per channel over any 20-second period in most instances). This is considered wideband communication because it takes place over a wide band of communication channels — 902–928 MHz.

See Chapter 5 for more details on understanding the physics of tags and choosing the right one. See Chapter 9 for all the specifics of testing for the proper tag and placement.

Part II

Ride the Electromagnetic Wave: The Physics of RFID

The 5th Wave   By Rich Tennant

"We're here to clean up the ambient electromagnetic noise."

In this part . . .

You start to take your first steps down the RFID road in this part. I give you more in-depth knowledge of the technology and its various uses. You explore some of the specific implementations and understand the specifics of how tags, readers, and antennas all work together as a system.

This part gets you up to full speed on all the workings of an RFID system and really goes a long way toward satisfying the geek in you. You find out about air interface protocols, details of antennas, and what makes a reader work.

Chapter 4

What Makes Up an RFID Network

*I*f this book is your first bit of reading on the topic of radio frequency identification, this chapter serves to ground you in the basics of radio frequency technology and how it came around to tracking small tags through the supply chain. If you've been inundated with press about the newest and most disruptive technology in decades, this chapter is your reality check. This chapter walks you through some of the basics of good old-fashioned physics and the laws that God created for science.

In this chapter, I describe the elements of an RFID network. The description starts with basic RFID components and moves step by step to a full Web-enabled supply chain network. This step-by-step approach helps you understand how a single tag the size of a quarter can grow into a global supply chain technology that is capable of saving billions of dollars. The RFID network is built up one node, and one tag, at a time. Understanding the individual components and how they fit together helps you frame architecture and deployment strategies.

Elements of a Basic RFID System

Learning the fundamentals of RFID can be overwhelming. You can avoid feeling overwhelmed and the sensation of going around and around in circles, and the sensation of going around and around in circles — I'm just messing with you — by understanding the basics of how data travels in waves and then through a network in an RFID system. This understanding gives you a solid foundation for greater knowledge as you explore the global architecture of RFID.

In a basic RFID system, four fundamental components are required for data to make its grand journey:

- **A transponder** (more commonly just called a *tag*) that is programmed with information that uniquely identifies itself, thus the concept of "automatic identification"

- **A transceiver** (more commonly called a *reader*) to handle radio communication through the antennas and pass tag information to the outside world

- **An antenna** attached to the reader to communicate with transponders

- **A reader interface layer,** or middleware, which compresses thousands of tag signals into a single identification and also acts as a conduit between the RFID hardware elements to the client's application software systems, such as inventory, accounts receivable, shipping, logistics, and so on

Figure 4-1 shows the basics of how a simple RFID system works and the four main components of that system.

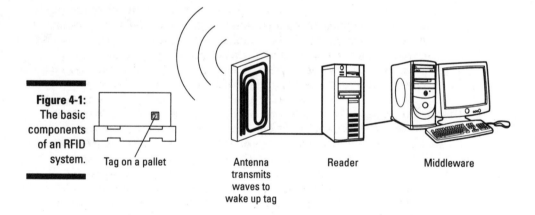

Figure 4-1: The basic components of an RFID system.

Tag on a pallet   Antenna transmits waves to wake up tag   Reader   Middleware

For now, I focus on a passive RFID system and a passive tag. Here's an overview of how the system works:

1. The tag is activated when it passes through a radio frequency field, which has been generated by an antenna and reader.

2. The tag sends out a programmed response.

3. The antenna that generated the field originally and is attached to the reader detects that response.

4. The transceiver (or reader) sends the data to the middleware.

5. The middleware sends the information contained in the tags to whatever systems need that information.

The following sections discuss the role each piece of the system plays in creating waves and transmitting data through those waves and into your network.

Everything starts with the tag

A tag, in a passive RFID system, is a little transceiver waiting to be turned on (and no, that doesn't happen by seeing a tag of the opposite sex). The tag has a small computer chip (or memory area) that is programmed with information that uniquely identifies the tag. This information is sent when the tag is activated (turned on).

A passive RFID transponder does not contain its own power source; rather, it absorbs energy propagated from a reader antenna's radio frequency (RF) field to supply all the power it needs to wake up its chip and communicate with a reader by sending back (backscattering) the information contained in its memory to a receiving antenna. As tags move into an antenna's radio field, they are excited, and each one transmits its identification data.

For more on the inner workings and physics of tags, see Chapter 5.

Antennas send and receive radio waves

Both tags and readers have their own antennas because they are both radio devices. A tag antenna, which is only a few centimeters (or less) long, attaches to the integrated circuit (IC, or just *chip*) to absorb a signal and then transmit out a slightly modified signal. The reader antennas range in size but are generally about the size of a computer flat screen and are specially tuned to transmit and receive RF signals.

Antennas are how readers communicate with the outside world. Reader antennas send radio signals into the air to activate a tag, listen for an echo (or *backscatter*) from the tag, read the data transmitted by a tag, and, in some cases, write data onto a tag. Antennas act as conduits between the tag and the transceiver and can function continuously or on demand.

- ✔ Continuously active antenna systems are used when tagged items are present on a regular basis or when multiple tags are passing through the antenna's detection field.

- ✔ On the other hand, an antenna's detection field can be activated only when needed by a sensor of some kind. The on-demand method can be triggered by optical, pressure, or other kinds of proximity sensors.

Antennas come in a variety of shapes and sizes; I explain how to choose the right one from a physics perspective in Chapter 5. This diversity in size and shape allows antenna placement in a wide variety of locations — from warehouse doors to highway tollbooths.

Readers tell the antennas what to do

An antenna is connected to a transceiver (which is generally known as a *reader*). Typically, one to four antennas are attached to a single reader, and those antennas send out the reader's signals. Basically, the reader tells the antennas how to generate the proper RF field, which can cover an area as small as 1 inch to as large as 100 feet or more, depending on the power output and the frequency. When an RFID transponder (or tag) moves into the antenna's radio field, it becomes active and sends back to the antenna whatever information has been programmed into its memory. A reader receives the tag's signal through its array of antennas, decodes the signal, and sends the information to the host computer system. A reader can also transmit special signals to a tag — telling a tag to come alive, synchronizing a tag with the reader, or interrogating all or part of the tag's contents.

The middleware transforms the system into a network of objects

The basic elements of an RFID system are rarely useful in isolation. They gain value as part of a production or logistics system. In this way, the use of more than one system in an industrial process becomes a local network. The connection of local networks constitutes a global network. You can think of the local networks as a node of hardware (a reader, antennas, and tags) that interacts within itself to exchange information over RF waves. A bunch of nodes connected together creates a global network that connects to an application that creates useful information out of the data.

In order to move data from a single node to the local network and/or to the global network, you need the data-collection component, which ties readers, antennas, and tags together. This component is called by many names — middleware, reader interface layer, Savant — all describing the very simple glue that sticks together each node in an RFID system.

Middleware connects the data coming into a reader to the client's host software systems. The middleware provides a coherent and stable interface between the RFID hardware operations and the flow of data elements, such as EPC (electronic product code) numbers, into inventory, sales, purchasing, marketing, and similar database systems distributed throughout an enterprise.

The elements of middleware include the following:

- ✔ **Reader and device management:** RFID middleware allows users to configure, monitor, deploy, and issue commands directly to readers through a common interface.

- ✔ **Data management:** As RFID middleware captures EPC data or other data from readers, it can intelligently filter and route it to the appropriate destinations.

- ✔ **Application integration:** RFID middleware solutions provide messaging, routing, and connectivity features required to integrate RFID data into existing supply-chain management (SCM), enterprise resource planning (ERP), warehouse management (WMS), or customer relationship management (CRM) systems.

- ✔ **Partner integration:** Middleware can provide collaborative solutions like business-to-business (B2B) integration between trading partners.

The basic elements provide the data source or the local node to generate data. A series of these are linked into a local network that can connect to either a larger network or even a global network by employing middleware. An RFID network is a peer-to-peer architecture capable of aggregating highly actionable data to a central location. See Chapter 10 for more details about middleware.

Imagine this: The use of a single tag, no larger than a book of matches, is multiplied millions of times over within a global supply chain, which creates a peer-to-peer network that shares data in real time across a limitless number of boundaries. The image of the single millimeter-sized chip quickly expands to comprise a warehouse; a company; an industry; and a world of rapidly changing, automatically updated, real-time information. From that tiny chip blossoms the power to know where every object is at all times in a global network. Pretty cool, huh?

Time to Make Some Waves — Electromagnetic Waves

To understand today's new RFID technology and equipment, it is important to understand the fundamentals of the science. RFID is all about physics. Laws and mathematical equations that describe the behavior of this technology have been around for decades, even centuries. Although some people might have you believe that a successful RFID deployment requires you to wear sacred shells, sacrifice a chicken, and walk across hot coals, black magic usually isn't required.

Plowing the fields for electromagnetic radiation: A timeline

As far back as science knows, various fields of electronic and magnetic radiation have existed. But the field of radio frequency communication didn't really take off, from the perspective of RFID, until the late 1800s:

- **In the 1860s,** while all his friends were out playing golf, James Clerk Maxwell, a Scottish physicist, predicted the existence of radio waves and postulated uses for those waves.

- **A short while later, in 1886,** German scientist Heinrich Rudolf Hertz skipped Oktoberfest to prove that rapid variations of electric current could be projected into space in the form of radio waves similar to light waves, and that this current was measurable and repeatable.

- **In 1902,** Italian physicist Guglielmo Marconi sparked a signal from England across the Atlantic to the shores of Newfoundland, demonstrating the first long-range use of radio waves as a form of communication. He broadcast the letter *S* in Morse code. He was trying to transmit *SOS* (with the classy Italian accent that worked so well for Sophia Loren) but left the folks in Newfoundland hanging.

- **During World War II,** the British developed the first RFID tagging system in order to rapidly discriminate between their own returning aircraft and squadrons of the German Luftwaffe. British fighters were equipped with tags that replied to an interrogation signal with a special "I am a friend" code — routinely changed so that the enemy could not use it. Snoopy should have had this in his battles against the Red Baron.

- **In the late 1960s,** the need for security and safety of nuclear materials drove further development of RFID tagging, such as electronic article surveillance (EAS).

- **In 1977,** Los Alamos Scientific Laboratories (LASL) transferred the RFID technology that had been developed in government labs to the public sector. Commercial RFID applications beyond EAS began to appear in the early 1980s: railroad freight car tagging, the tagging of cattle and rare dog breeds, automobile immobilizers, keyless entry systems, and automatic highway toll collection.

- **As the 1980s drew to a close,** the primary focus in RFID commercialization shifted from new applications to issues of performance improvement and cost reduction, as well as reader, tag, and antenna miniaturization. The success is evident in the variety of RFID applications and system components now available in stores like Radio Shack.

The best RFID engineers understand where the technology originated, what its limitations are, and how the laws of physics can be leveraged as an asset in the design and deployment of an RFID network. The following sections explain basic principles of electromagnetic waves, how they're measured, and how they affect each other.

Understanding everything that happens in the environment is critical to the success of your RFID system; that's why we spend all of Chapter 7 talking about the site assessment. But knowing the root cause of problems will help you fix issues that crop up during that assessment.

Your "new" experiences are grounded in history. Problems you are confronting for the first time have likely been solved before. I always say it's great to learn from mistakes — as long as they are someone else's.

Frequency is a measurement

Electromagnetic radiation may have begun when the quark and leptronic soup became transparent to photons (the electromagnetic carrier particle), but the ability to measure all that radiation arrived relatively recently in the 19th century. Scientists like James Joule and James Maxwell were the first to figure out that you can measure the invisible. The ability to quantify frequency began with the advent of modern physics and the development of wave and particle theories. It is the principles behind wave and particle theory that led to the use of electronic and magnetic waves to communicate data.

Frequency is an important topic in the understanding of RFID. In Chapter 3, I introduce the three main types of frequency in RFID: low frequency (LF), high frequency (HF), and ultrahigh frequency (UHF). As you begin to understand the physics of RFID, you need to understand how frequency works as a measurement:

- *Frequency* is a measure of how many times an electromagnetic wave goes from one crest to the next crest in a unit of time (such as a second) as it moves through space.

- This movement from crest to crest (or trough to trough) is called a *cycle*. Frequency is measured in Hertz (Hz), which tells you how many cycles per second occur in an electromagnetic wave.

When ultraviolet radiation burns our skin at the beach, X-rays take pictures of broken bones, light glows from a neon sign outside your hotel window, or signals are sent to antenna arrays in RFID networks, different frequencies are at work. All these sources belong to the family called the *electromagnetic spectrum.* As the name *spectrum* suggests, radio-frequency emissions form a series starting at extremely low frequencies (such as your car radio), going through the familiar visible wavelengths at higher frequencies, and finally to X-rays, gamma rays, and cosmic rays at extremely high frequencies. For example, the visible region of the spectrum is around 10^{14} Hz, and the UV rays that burn us are at 10^{16} Hz. Cosmic rays are 10^{21} Hertz. RFID normally uses a portion of this spectrum from 125 kHz (kilohertz) to 5.3 GHz (gigahertz).

History may repeat itself, but virginity comes only once

In 1902, when the Marconi team sent out the first transatlantic RF signal, that sole signal had the world's airwaves to itself. Lightning was the only possible competition. Fortunately, the skies over the northern Atlantic were clear that day, and long-range radio was born. That signal was received 1,500 miles away because no competing RF signals were around to create interference.

In the late 1990s, when modern RF engineers from the Marconi Corporation (named after the famed engineer) proposed a reenactment of the original event, it turned out to be impossible due to RF crowding or *noise.* A small signal discharged today is completely drowned in the vast sea of radio frequency noise. The lesson here is that the human eye couldn't detect a change in the radio wave patterns in the last hundred years; the invisible noise is what creates the biggest stumbling blocks to any radio system.

You need to be aware of all RF sources in and around your environment and also of any other sources that might interfere with the RF transmission once your network is operational. I show you how to check this in Chapter 7.

Fields: Electrical and magnetic, near and far

An electrical current gives rise to a surrounding magnetic field. A common example of the effect of this field is the effect of a current on a compass needle: The electrical current generates a magnetic field, which causes the compass needle to align with the magnetic field. A magnetic field can also generate an electrical current. This dual relationship between electrical and magnetic fields is a basic and fundamental physical property.

The region close to the source of the electrical current, where the magnetic or electrostatic forces can be detected, is called the *induction field.* Outside the induction field is the *radiation field.* Depending on which type of frequency your system uses (LF, HF, or UHF, which I introduce in Chapter 3), either the induction field or the radiation field will power the tags:

- ✔ **In LF and HF systems,** the induction field has sufficient power to drive an electromagnetic field in the tag so that the chip is activated. Outside the induction field, the radiation field is too weak to do the same to other chips. This means a reader won't activate tags in neighboring LF or HF systems.

If you're interested in why the induction field can power the tag, research the *inverse square law*.

Measuring the strength necessary to actually activate a tag in the induction field is how you focus your RFID system's detection areas. This detection distance in the induction field is called the *near field*.

✔ **In UHF systems,** the radiation field powers up the tag. This detection distance is known as the *far field*. Because you're working in the far field, the antennas are shaped and work differently than antennas in LF and HF systems (more on antennas in the next section).

Creating resonance between the antennas and the field

Antennas are made of conductive material and *couple* the RF waves for communication purposes. *Coupling* is the matching of the tag and the reader so that they can communicate effectively together at the same frequency. Every piece of electrically conductive material has some degree of coupling with radiation fields out in the real world. Only when the conductor is designed to provide high coupling efficiency between certain media is it called an *antenna*.

A key feature of antenna design is the idea of a *resonance frequency*. *Resonance* means that two things are moving in unison or in lock step. Ignoring for the moment the underlying mathematics and physics of this event, you can easily demonstrate how resonance works. Fill a long, low basin with water. If you put your hand into the water and, with large strokes, move it back and forth, the water becomes turbulent and you splash water out of the basin. But, if you gradually change the length and frequency of your strokes, you will eventually find a rate at which the entire body of water moves in unison with your hand. This is the resonance frequency. Your hand has matched the resonance frequency of the water in a basin of those particular dimensions.

Antennas work the same way. They need to match the frequency of the incoming field in order to set up a resonance between the antenna and the field. Resonance is based on a multiple of the wavelengths; thus you will notice that tags (which are tiny transceivers with their own antennas) have a size that is generally proportional to the size of the reader's antenna.

The shape of the antenna is also matched to the frequency it is intended to interact with. Thus LF and HF tags are shaped like coils, which resonate better in the near field, and UHF tags have a flatter shape, which works better in the far field. The simplest antenna design of this nature is an antenna called a *half-wave dipole antenna* (a good term to remember for cocktail parties when someone asks you about RFID). Essentially, the idea is to match half of the wavelength (half wave) with the resonant frequency, and it will receive a stronger signal.

It makes sense to relate the physical concepts of resonance to some practical RFID issues:

- ✔ When a tag antenna is immersed in the field of a reader antenna (in which both antennas are tuned to couple at the same frequency), the tag absorbs the radio frequency energy at a particular wavelength — the wavelength that makes it move at the same rate as the reader antenna. This is how resonance works in an RFID system.

- ✔ The UHF antenna design is proportional to about the wavelength of the signal. Knowing the wavelength of UHF systems (about 33 cm) is important when designing your RFID system because anything conductive that is about that length can act as an antenna and cause problems with your system. Tags that are some multiple of that wavelength will also receive a better signal.

ODIN technologies' top engineer was doing a site assessment at a company that made cable racks for data centers, among other products. The engineer was trying to figure out why all the RF from a signal generator was going haywire in a particular part of the warehouse. After looking around, he noticed that the metal ladder racks for cabling were made with a bunch of 1-x-1-foot sections. He realized that 1 foot is about 33 centimeters, which is a perfect wavelength for UHF. The ladder racks were absorbing all the RF signals intended for RFID tags and needed to be relocated to make the system work properly.

Chapter 5

Understanding How Technology Becomes a Working System

*I*f you want to know a thing or two about setting up a great RFID system, take a look at Lance Armstrong, one of the greatest athletes of all time. What does Lance have to do with RFID? Well, you can make even the simplest and most well-proven systems better by paying painstaking attention to the minutest details. You also need to understand every component that makes up the entire system. The simple system in Lance's case is a bicycle. If you add a little oil to a rusty chain, an old neglected bicycle can take you around the neighborhood or through the countryside in far less time than you can walk. Lance's team in the Tour de France, by contrast, has spent countless hours understanding the effects of wind, speed, tire pressure, clothing, and so on, using that knowledge to fine-tune their clothing and bicycles so they don't lose a second to the competition. Sure, the cyclists look ridiculous in their pointy helmets, but their efforts do illustrate the benefit of understanding every component of a system, how the system functions, and the effect that the system can have on overall performance.

In this chapter, you come one step closer to crafting the Tour de France bicycle of RFID. I walk you through the basics of RFID tags and readers, talk to you about different aspects of design and performance, and show how they interoperate. After you understand the technology, I take you through a crash course in understanding the protocols that allow tags to communicate with readers.

With a more in-depth understanding of the individual subcomponents of an RFID system, you'll be much better able to create a high-performing, efficient RFID network. If you understand how the system works on the basic level, you can begin to investigate specific ways of doing things in your environment, and begin to influence what technology, protocols, and standards you wish to support.

Anatomy of a Passive Tag: Understanding How It Works and Choosing the Right One

All the buzz is about passive tags. Thanks to the work of the Auto-ID Lab at MIT, passive RFID tags have become cheaper, smaller, and globally interoperable (for the most part). See Chapter 21 for more info on the standards. Although your RFID applications might require active or semi-passive tags (Chapter 2 can help you determine which type you need), the innovations in passive tags drive the Wal-Mart and Department of Defense (DoD) mandates, and this is the tag that you're likely to use.

Although a tag seems quite simple, like a bicycle — it's just a couple of wheels and gears and a frame, right? — the engineering and innovation surrounding a tag can make all the difference in the world. Understanding how tags work and some of the basics behind tag design will help you to understand the potential applications and types of passive tags that you might want to investigate. The two parts of a tag that have the greatest fundamental impact on its performance are the tag antenna and the integrated circuit (IC, or chip). In this section, I explain how these two components are responsible for extracting, efficiently consuming, and reflecting RF power.

As you consider how the different tag designs work and which design will work best for you, think about where you might put the tag on your object, case, or item as well as how the tag might be attached to the item in the manufacturing, assembly, or packaging processes. Having an idea where the tag will be enables you to choose the *optimal* tag design — that is, one that enables you to get accurate reads. In Chapter 9, I explain how you place tags and test them in a real-world environment, but for now, it's helpful to keep tag placement in the back of your mind.

How do tags receive and transmit information?

Think of an RFID tag as a very simple, very small two-way radio. Depending on the type of tag, its role is to receive a signal, power up an onboard chip (with data embedded on it), and then broadcast that information. It is a simple yet elegant bit of science at work right under your eyes. Well-known laws of physics govern the tag behavior, like they govern the behavior of a bicycle balanced on two wheels.

A passive tag consists of three components that enable it to receive and transmit signals in its ultra-efficient way:

- ✔ **An integrated circuit or chip:** This chip stores data and executes specific commands. Most of the passive tags today carry 96 bits of memory, although some can carry as little as 2 bits or as much as 1,000 bits. The chip design determines whether the tag has read-only or read-write properties. Three or four manufacturers, including Phillips, Texas Instruments, Fairchild, and ST Micro, are the primary makers of these chips. Most tag manufacturers purchase their IC chips from these primary sources, so you have little chance of getting a cost advantage in chip purchase.

- ✔ **An antenna (or coupling element):** The antenna's function is to absorb RF waves and then broadcast a signal back out. The antenna powers up the tag by collecting the energy from the RF field and exciting the onboard chip into action. This process is called *coupling* because the tag antenna must "couple" with the electromagnetic fields that the RFID reader emits. In more technical terms, coupling describes the extent to which power is transferred from one system component to another — in this case, from air to the antenna.

   The size of the antenna is critical to the performance of the tag because the antenna's size usually determines a tag's read range. Simply put, a bigger antenna can collect more energy and therefore broadcast more energy back out. You can usually determine the size of a tag's antenna by looking at the tag: the bigger the tag, the bigger the antenna. You might see antennas (and therefore tags) that are the size of a dime and others that are the size of a business envelope.

   Another characteristic of antennas that enable them to send and receive signals is the antenna's shape: Low-frequency (LF) and high-frequency (HF) antennas tend to be coils because these frequencies are predominantly magnetic in nature. Ultrahigh-frequency (UHF) designs tend to look like radio or old-style television antennas because UHF frequencies are more purely electric in nature.

   Later in this chapter, I explain how tag antennas work and how different engineering designs can improve antenna performance in various RFID applications.

- ✔ **The *substrate* (or material that holds it all together):** This is most often a Mylar or plastic film. Both the antenna and chip are attached to it.

Innovations that might save you money

Passive tags made widespread adoption of RFID affordable, and innovations in engineering and production processes will help make tags even cheaper.

Conductive ink is an area of antenna innovation that promises to drive down the costs of RFID tags significantly. *Conductive ink* is essentially ink with properties that are amenable to RF *coupling* (connecting the broadcasting signal to the receiving tag in an optimal manner). The benefit of using conductive ink to make the antenna covers both material costs and engineering costs. Most traditional metallic antennas are made by taking a solid piece of metal, often copper, and removing material to get to the desired shape. Obviously, this wastes a lot of good copper or aluminum. Conductive ink, on the other hand, uses various printing technologies similar to inkjet printers that *add* only the amount of antenna material needed, making it much more efficient.

Another area for innovation and cost savings is the substrate that holds tags together. Many people tend to overlook the significant expense of traditional chip adhesive and assembly processes. Historically, chips have been attached by using a *flip-chip* (flipping a chip into place and gluing it to an antenna) assembly process, which is not only costly (when you're talking about fractions of a penny for components) but also slow. Two unique innovations in the tag manufacturing process have been Alien Technology's Fluidic Self-Assembly (FSA) and Matrics/Symbols Parallel Integrated Chip Assembly (PICA). Both hold promise to dramatically reduce production cost and speed up capacity.

How does a tag antenna work, and how do you choose among the different kinds?

Tag antennas come in all shapes and sizes, and the antenna design changes things dramatically. A wide variety of antenna designs has been proposed in attempts to maximize the performance of the tag on a wide variety of materials.

The fundamental problem of RFID is transmitting adequate power to RFID tags. The transmitting power is considered adequate when a tag can efficiently consume, use, and reflect RF power when attached to a case or pallet (usually this requires –10 db of power). Understanding how different tag antennas work — and especially how they reflect power back (a process called *backscatter*) — helps you make the right selection and ultimately leads to optimal performance.

For an RF wave to properly power up a passive tag, the electrical current coming out of an RFID reader has to hit the conducting plane (the antenna) *orthogonally* — that is, at right angles. This simple law of physics, known as Gauss's Law, states that electric flux creates a charge and that an electric field cannot just go past a conductor — it must turn and meet it at right angles. So what does this mean to practical design application? Here are some points to keep in mind:

- ✔ **Antennas that have many different angles are designed to couple with an RF wave at any opportunity.** That's why some of the antennas have many turns and wings shooting off the center. These antennas, which are called *orientation-insensitive,* are better for reading a tag as it passes through a dock door or doorway, for instance.

- ✔ **The long, straight tags, on the other hand, are designed to perform very well on flat, directionally sensitive applications or with a circularly polarized antenna.** You can use these to good effect on cases going down a conveyor belt. The tag reader signal comes from a constant, predetermined direction. Thus, with a little planning, the readers can hit the sweet spot with a whole lot of antenna area.

- ✔ **The straighter the tag antenna, the greater the size of the conductive plane (or coupling element); the greater the conductive plane, the better the tag performance.** If the tag's antenna is curved in many directions, only part of the tag is ever orthogonal to the RF wave, so only part of the antenna is used. If the tag has a straight antenna and the antenna is in proper orientation, the entire surface becomes used in power and communication. That means the tag has a greater read distance and is more likely to receive the power needed for accurate reads.

With the proper understanding of tag antenna physics under your belt, you can ask the following questions to determine whether a given tag antenna design is right for you:

- ✔ **What are the coupling characteristics of the antenna?** All tag antennas have a *capacitive* element (a plate to store magnetic energy) and an *inductive* element (the coil to store electric energy), which make up the *impedance* (how easily current can flow through a system, measured in ohms) of the antenna.

Some tags are tuned: Just as a tuning fork is tuned to a particular key, an antenna can be tuned to a particular frequency specifically to work best when affixed to cases of product consisting of metals, liquids, or other specific materials. The length of the antenna determines the tuning, as shown in Figure 5-1.

Figure 5-1:
The length of the antenna determines the receiving frequency.

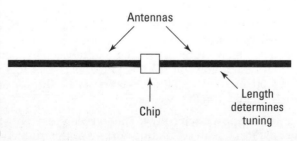

Antennas

Chip

Length determines tuning

Others tags actually use the product as the antenna. For example, several manufacturers are designing tags that couple with a DVD, using it as the antenna.

✔ **What is the orientation sensitivity of the antenna?** As I mention earlier in this section, some antennas read from many different directions, and others read from just one.

Some tag designs effectively incorporate multiple antennas, each of which is polarized in a different direction but in the same plane. This allows you to put antennas on in multiple directions and still get good reads as opposed to affixing them in a specific orientation to be oriented specifically to a reader's antenna. Another innovative idea in tags is to actually put two antennas on a chip, which is often called a *dual-dipole tag.* This idea gives you twice the orientation sensitivity because the tag antennas are usually mounted at right angles to each other on the tag.

✔ **How does the tag fit on the product?** Some tags achieve superior range and orientation insensitivity at the expense of small form factor. If these tags do not fit in the space allocated for labeling purposes, they should be eliminated from stock-keeping unit (SKU) testing (which I discuss in Chapter 9). In other words, some tags might work great for a particular type of product, but if they don't fit on the product, they're not the right tags.

How does the integrated circuit affect performance?

Most tags are identified primarily by their antenna shape, but a microscopic integrated circuit has a far greater impact on overall tag performance. The IC, or simply *the chip,* is responsible for converting RF energy into usable electrical power, storing and retrieving data, and modulating the *backscatter signal* (the signal that the tag sends back to the reader). Tag parameters related to power extraction, consumption, and reflection include

✔ **The amount of memory on the chip:** Because low cost is the ultimate design priority of the electronic product code (EPC) industry, memory storage levels are kept to a bare minimum (96 bits on average). Rather than store all the data about an item in the tag's chip, the EPC uses a serialized numbering system to point toward additional information about each item, which is stored on a secure database. As such, the power required for encoding and reading EPC tags is kept to a minimum, on the order of 100 micro-watts ($1 \times 10E-6$ W) or –10 dBm. See Chapter 2 for more details on how the EPC works.

✔ **The efficiency of the power circuitry:** The IC receives energy from the tag antenna in the form of an oscillating current at the frequency of the reader transmission. This current must be down-converted and rectified by using circuitry tuned to a specific frequency. The precision of these components and how well they are matched determine power conversion efficiency. Some newer chips from companies like Impinj are being designed to operate more efficiently and thus use lower power than traditional chip design, which would mean that your readers could have lower power output and less interference with each other.

✔ **The impedance match of the chip and the antenna:** If an impedance mismatch exists between the chip and the tag antenna, power is reflected away from the chip and thus unavailable for use by the tag. This is the case with some poorly manufactured tags. Unfortunately, the only way to find out this for sure is to use an expensive piece of test equipment called a *network analyzer.* The better way to avoid this is to find historical data about tag quality by looking at commercially available benchmark studies.

✔ **The ability of the chip to alter the impedance of its antenna:** Tags send a signal back to the antenna by using a technology called *backscatter.* That backscatter can also *modulate* (change the signal) as the chip alters the *impedance* of the tag antenna (changes the ability of current to flow through it) at specific time intervals (pulse-width modulation). Think of this as taking a garden hose and squeezing it at a specific interval to see how the water changes coming out. The chip's ability to change the impedance precisely and in sync with the reader determines signal clarity and strength.

✔ **How tags must respond in collision-free channels:** When multiple tags pass through the RFID reader's field simultaneously, they must talk in turn to prevent data collision at the receiver. EPC tags support one of two algorithms used to accomplish this task: tree walking and ALOHA slot. The anticollision protocol determines performance, although the emerging standards will likely set the protocol to be an ALOHA slot. This means that if you want the slightly better performance of a tree walking protocol, you need to use a proprietary system.

The mu-chip takes over the world (when cows fly)

Recently, Hitachi, Ltd., introduced the *mu-chip*, an RFID chip and antenna on an IC board about the size of a pinhead. The press had a field day with the release of the product, saying that it would now be possible to clandestinely track everything all the time.

If you understand the relationship between antenna size and tag performance, you know the physics of RF communication make this Orwellian scenario impossible. The maximum read range of the mu-chip is just a few millimeters because the limited surface area of the antenna limits the chip's power and therefore its ability to transmit data off the chip. The chip will have great utility for anti-counterfeit purposes in everything from money to concert tickets, but like the other laws of physics, you can't violate the principles of antenna size no matter how felonious you might be feeling.

Each IC manufacturer has a proprietary chip design that employs a unique manufacturing process. The ability of the manufacturer to optimize each of these parameters will determine in large part how well the tag performs.

Some tag examples for the geek in you

Tag antenna designs are a combination of art and science. Many tag antennas are designed with sophisticated computer modeling programs, and others are designed by engineers, using known shapes and patterns from other applications. You can see how the tag designs vary in Figure 5-2.

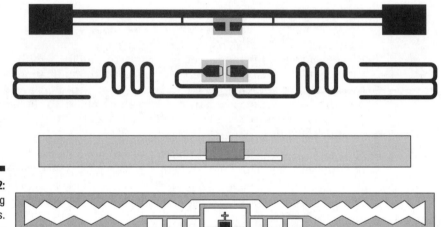

Figure 5-2:
Various tag designs.

Some of the more popular passive tags being used in the Wal-Mart and DoD deployments include the following:

- ✔ **Alien "I2" tag:** The Alien "I2" tag has an advantage over other tags in that its length approaches half a wavelength (approximately six inches) at 915 MHz — the ideal length of a dipole antenna. It exhibits a very high level of performance, particularly when mounted parallel to an antenna's field. **Dimensions:** 6.0 x 0.65 inches

- ✔ **Alien "Squiggle" tag:** This tag "squiggles" in two dimensions to gain virtual antenna length, making the orientation and length of the antenna element optimized while keeping the tag compact. Thus, if the tag is not in the perfect orientation, it still has the chance to couple with the broadcasting antenna. **Dimensions:** 3.8 x 0.6 inches

✔ **Avery Dennison Strip tag:** The Avery Dennison Strip tag is unique in that it is nearly all metal, like a normal transmitter antenna dipole, which allows it to have a more conductive surface and absorb more energy. **Dimensions:** 3.75 x 0.45 inches

✔ **Rafsec Folded Dipole CCT tag:** The Rafsec tag is unique because it is a folded dipole. The current for the antenna is strongest at the midpoint of the antenna; accordingly, the strongest radiation occurs at the center of the antenna substrate, along the upper strip. This offers good long-distance read range. **Dimensions:** 4.0 x 0.5 inches

These four tags illustrate just a few of the possibilities out there. In fact, several dozen different tag types are available from manufacturers like Alien, Symbol, OMRON, Rafsec, Avery Dennison, Texas Instruments, and others. If, as you investigate and test tags, you don't find a tag that works for you, consider companies that do custom tag design (using anything from advanced fractal mathematics and sophisticated programs to geometric shapes from an artist's mind).

Tracking the Tags with a Reader

Before you decide to quit your day job and open up a tag design boutique, you'd better add reader functionality to your list of growing knowledge. No matter how sophisticated a tag is, it's worthless without a reader. A reader is an information tollbooth on the highway to efficient supply chains, accurate inventories, and perfect asset management. That is, readers collect the important information from the tags as the tags pass through the supply chain applications so you can make useful business decisions based on real-time information, like when to order more stock.

Understanding how readers work will help you understand the system better and ground your knowledge for assessing the various types of readers.

Holler back, young 'un — Transmitting and receiving signals

An RFID reader is a sophisticated radio. To illustrate how a reader works, the following steps walk you through the life cycle of a read:

1. The energy to transmit the radio wave comes from an external power source like a battery or a wall outlet.

2. Inside the reader, a digital signal processor (DSP) chip and a regular processor control the flow of electricity in a very specific manner, modulating the frequency and the amplitude of the wave that the reader generates.

I explain how this works in more detail in a moment.

3. That flow of electricity goes to an antenna via a coax cable. How the electricity gets to that antenna is controlled by the complex circuitry anchored by the DSP.

4. The antenna sends out an RF wave carrying data by using a process called *modulation*.

   Modulation is essentially the introduction of very small variations in the electrical signal. An easy way to picture this is by imagining the old Civil War signalmen who flashed light signals back and forth from ship to ship or from ship to shore. That was wave modulation, but they were modulating light waves. Of course, a sophisticated RFID unit uses a much more complex mechanism in which the frequency and/or amplitude of the transmitted wave are varied in the slightest manner to encode a great deal of information. Whereas the Civil War signalmen transmitted dots and dashes to remind someone to bring more rum back to the boat, the reader's RF signal transmits 0s and 1s that an application turns into information about that item.

5. After the reader antenna receives the signal back from a tag, the reader carries the signal back down to the electronics.

6. The electronics then make sense of the subtle differences in the waves and decode it to create useful information. (Note that this is different from the filtering and smoothing that middleware does on the data or EPC numbers received; see Chapter 11 for more on middleware.)

Transmitting antennas are represented by the abbreviation *Tx,* and receiving antennas are represented you guessed it — *Rx.* In many cases, they are the same antenna; however, on some readers, you might see a spot to plug in a Tx antenna and another one to plug in an Rx antenna. The Tx antenna is the one broadcasting a powerful signal, and the Rx is the one listening for the much weaker signal from the tag. If the Tx and Rx antennas are separate, you always want the Rx signal as close to the tag as possible.

The DSP chip: Examining the brain of a reader

As I mention earlier in this chapter, the digital signal processor (DSP chip) controls the electricity that flows through a reader. Specifically, it applies an alternating voltage (for example, the modulated carrier wave carrying the information) to a transmitting antenna. This process of producing a current that moves back and forth (*oscillates*) is more complex than it sounds because the number of oscillations over a period of time determines the information on those waves. All this oscillation and frequency generation requires the DSP to do a lot of sophisticated math.

In addition to the DSP, every RFID reader has a fairly standard onboard processor to do simple calculation and run the operating system. Figure 5-3 shows the inside of an RFID reader and points out the DSP and the primary processor. As you compare various readers and their technology, you'll want to know who makes their DSP and main processors, particularly because many people are going to rush to order RFID readers as they are required to. This immediate order will cause production strains on boutique makers of chips, whereas companies like Texas Instruments and Intel are more likely to be able to handle huge production volumes.

The DSP, the heart of any RFID reader, has four specific properties:

- ✔ **Mathematical whiz chip:** The DSP chip is first and foremost a calculator on steroids — the basic design leverages arithmetic logic units and one or more multipliers in their primary function. These processing units are designed to be extremely fast and to execute, in a single *clock cycle* (one unit of time for a computers processor to run), the full extent of their mathematical operations.

- ✔ **Super-efficient memory:** Programming on DSP chips needs to be highly efficient because there is such a limited amount of memory. The average DSP chip holds anywhere from 8 kilobytes to 256 kilobytes. To put it in perspective, this chapter is about 60 kilobytes as a Word file (with no graphics).

- ✔ **The ability to move data in and out in real time:** A DSP chip is the ultimate inventory management device — it gets data in and data back out in a real-time continuous stream. It has to, or else a gap in the communication occurs because radio waves can't be stored in a cache anywhere. A lot of folks bandy about the term *real-time* when they actually mean *near* real-time. But in DSP processing, it does actually happen in real-time as a stream of constant processing of the electronic signals.

- ✔ **Low power requirements:** The DSP chip is like the Toyota Prius of the processor world, and the Pentium IV is like a Hummer. DSP chips use only a fraction of the power of a normal processor, even at full speed. Because they use less electricity to run, they also produce less heat. The Pentium IV processor, on the other hand, has incredible performance but uses up a lot of fuel to get that strong performance.

DSP chips are not just in RFID readers. They are the key to every electronic device that requires a lot of heavy lifting in the math department — from cellular telephones to MP3 players to digital cameras.

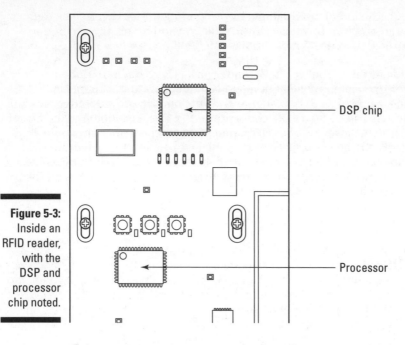

DSP chip

Processor

Figure 5-3:
Inside an
RFID reader,
with the
DSP and
processor
chip noted.

Ring around the dipole and other bad antenna stories

An RFID reader uses one of several basic types of antenna. As with any radio transmitter and receiver, the antenna is the ears and mouth of the system. It both "talks" (transmitter) and "listens" (receiver). For an RFID reader, you need an antenna with the following properties:

- **Highest directivity:** The antenna property of directivity is the guiding parameter for point-to-point communication, as opposed to broadcast communication. For example, compare a broadcasting application like FM radio with a wireless data network. In FM radio, the more people who can hear your station the better, so your goal is to spread your transmitter signal in all directions as far as it will go. But in a wireless data network that connects two buildings, you want the signal to radiate in a preferred direction so that not many antennas can receive the data. In the case of RFID, you want your transmitted signal and your received signal to be confined to a specific area, not broadcast into other RFID zones or systems.

Receiving signals from another RFID interrogation zone creates what are called *phantom reads* or *ghost reads*. Many people setting up systems via trial and error end up experiencing phantom read problems.

✔ **Best gain:** *Gain* is a measure of how much better (or worse) an antenna can transmit its signal over some known baseline reference. To help visualize this, an FM radio station wants its transmitting signal to be spread out in a full circle so that more people can hear it. The wireless data network administrators, on the other hand, want the transmitted signal to prefer one point to another. If you add all the energy in the circle and then take that energy and fit it into a highly directive antenna, the signal strength will be concentrated. In this case, the signal strength is much stronger at some reference point for the highly directive antenna than for the broadcast circle antenna. In RFID, you want the transmitted signal to be concentrated at the tag so that the tag will wake up quickly and backscatter a strong signal to the reader. On the receiver side of things, you want your receiver to be able to hear the smallest possible backscattered signal.

✔ **Circular polarization:** *Polarization* for an antenna is simply the orientation that the electromagnetic wave radiates from the antenna. The two basic types of polarization are *linear* and *circular.* In the simplest of terms, if your antenna is a dipole (or "stick"), it broadcasts out in a line that generates from that stick. With circular polarization, the radiating signal spins.

In any radio communication system, you need to have the receiving and transmitting antennas in the same orientation for maximum signal transfer from transmitter to receiver. In this way, the talker is loud, and the listener hears most efficiently.

With RFID and polarization in mind, you must look at the two radio links involved. The reader serves as the transmitter for the *forward link* (reader-to-tag) and also as the receiver for the *back link* (tag-to-reader). In a similar manner, the tag also operates as both a transmitter and receiver. For the most efficient communication between tag and reader, the orientation (or polarization) of the tag antenna and the reader antenna needs to be the same. If the orientation of your tags will be random, you can't use a linearly polarized antenna without losing read efficiency. For this application, an antenna with a circular polarization works best. When the tag is in any circular orientation with respect to the reader antenna, the reader will "hear" the tag well.

Other antenna parameters, such as *radiation efficiency* (the ability of the antenna to "push" a signal into the air) and driving point impedance, must also be known in order to design a transmitter that operates most effectively with the antenna. However, these parameters are usually fixed by the manufacturer to the radio industry standards. I won't clutter your head with an explanation that would be of no use. It suffices to say that commercially available antennas are designed correctly.

Air in Her Face — Blowing Sweet Nothings

That's *air interface,* not *air in her face!* The *air interface* (or Physical layer for you networking types) is the protocol that dictates how the readers and tags talk to each other and how the data keeps from colliding. The air interface protocol is the Cap'n Crunch decoder ring of the RFID world. A reader has a very specific way in which it encodes data by modulation, and this is the secret of the air interface. Without knowing how the information from the reader is encoded, the tag can't communicate with it.

Although RFID received much notoriety in the early part of this century, it wasn't until a well-accepted air interface protocol was designed that people began making significant investments in the technology. The air interface protocol (remember this because it's great cocktail party banter) that was finally chosen, after a year of deliberation, was the modified ALOHA slot protocol. This is what you need to make sure that your tags employ if you want to be compliant with EPC Generation 2.0 protocols.

Air interface technology

Here's a more detailed explanation of how anti-collision works within air interface technology:

Space Domains

In *space-domain methods,* tags are placed in specific locations to achieve isolation. Tags in space-domain methods are identified by variation of reader range (variation of power is transferred to passive tags) and/or by using directional antennas. RFID using only space-domain methods drastically hinders the effectiveness of the technology. Space-domain methods rely on the number of tags in a reader's range — if too many tags are in the area, collision results and reduces the reader's ability to interpret any signals.

Frequency Domains

Frequency-domain anticollision methods allow for robust wireless communications but can add excessive complexity and cost to an RFID system. Frequency Division Multiple Access (FDMA) systems divide the total available bandwidth into fixed-width channels. FDMA is costly because it requires accurate frequency sources and band-pass filters. Code Division Multiple Access (CDMA) systems have many advantages over FDMA systems. CDMA offers better adaptability to varying traffic load, increased capacity to read tags, and processing gain. CDMA and other spread spectrum (SS) methods are currently difficult and costly to implement for RFID systems but are popular in cellular telephony.

Time Domains

Most RFID anticollision methods are time-domain. In these methods, fractional communications from tags are varied in time. Time-domain methods can be classified into synchronous and asynchronous schemes.

- ✔ **Synchronous schemes** are those in which a reader transmits a query to a specific tag by using its UID (unique ID) number. This is an effective anticollision method because tags do not have to take turns communicating to the reader, and tags do not have to rely on a complete "uncollided" transmission to be identified. A reader can poll through a list of tags, but the polling method, also known as *tree walking* or *binary tree,* is relatively time-consuming and depends on the tag's UID number being known. Binary tree searches use binary code (groups of 0s and 1s) for communication and basically involve the reader actively sending a signal of a 1 or 0 to a tag. If the reader sends the correct number, the tag acknowledges it by transmitting the signal back to the reader. If the reader sends the wrong signal, the tag mutes itself and awaits another signal. Eventually, the computer deciphers the code of the tag. Symbol Class 0 tags use this anticollision algorithm, and ODIN technologies lab research has shown it to be an effective method for interrogating multiple tags quickly.

- ✔ *Asynchronous schemes* are those in which tags in the reader's field respond at randomly generated times. This helps to reduce the chance of collisions. The ALOHA scheme is asynchronous and involves a node transmitting a data packet after receiving a data packet. If a collision occurs, a node becomes saturated and transmits the packet again after a random delay. The reader transmits continuously until a collision does not occur. A slotted ALOHA transmission is performed in slotted times by making small restrictions in the transmission freedom of individual data packets. When packets collide under slotted ALOHA protocol, they overlap completely instead of partially, and this significantly increases the efficiency of data transfer.

ALOHA is more easily understood by using an analogy. For the purposes of the analogy, imagine 50 tags in an interrogation zone. Then imagine 50 railway stations as tags, a railroad car as a tag's transmission, and two tunnels that represent the antenna/reader. A train gets loaded with supplies at its respective station (which is representative of the data being delivered). After supplies are delivered through the tunnel, they are no longer needed, and thus the next train does not need to leave the station. One tunnel (the reader output tunnel) has trains coming out on one track, one behind the other, at constant speeds (this represents the electromagnetic waves being emitted by the antenna). At the 50 stations, the trains stop simultaneously, pick up their supplies (which takes a random amount of time, analogous to the random delay of ALOHA protocol), and take off for the next tunnel. The tracks all converge at that next tunnel, however, and the trains cannot touch each other while merging, or the "collision" will destroy them. Thus, no supplies will reach the other side (no data will be transferred to the reader). Trains will continue to leave stations until the supplies are delivered, and the station, receiving word from the "antenna/reader" tunnel, shuts down (representing a tag shutting off, or muting itself). The trains are long, and even though they move extremely fast, they sometimes overlap by random distances. The more trains there are leaving at random times, the more chance for collision.

This is representative of pure ALOHA protocol. In slotted ALOHA protocol, trains are allowed to leave only at certain times *(slots).* Imagine that the stations have stoplights, and that they all turn green and red at intervals just long enough to prevent partial train collision. If a train is ready to go at green, it goes; but if it is not, it must wait for the next light. This ensures that if two trains (signals) go at the same time, they completely collide. If they overlapped (as they do with pure ALOHA), the lagging train would still have the ability to collide with a train behind it. Now imagine that the stoplight flickers on and off multiple times per second, and that the trains instantly accelerate to light speed. It might be easier to see how slotted ALOHA is superior to pure ALOHA. As stated before, after the train has delivered its supplies, it sends a signal so that no train leaves its respective station, thus ensuring a smaller chance of collision. The newest EPC protocol, EPC Gen 2.0, is a slotted ALOHA protocol.

Chapter 6

Seeing Different RFID Systems at Work

. . . .

In This Chapter

▶ Setting up RFID applications at the dock door, conveyor, shelf, and so on

▶ Examining real-life RFID systems at work

▶ Discovering other areas where you can apply RFID

. . . .

*T*ime to put on your right-brain hat and think creatively about RFID. After you have a little bit of an idea what goes into an RFID system, you're ready to think about where you might be able to put RFID to use and how it can improve your business operations.

This chapter introduces you to some basic applications of the technology, both the ones you've heard of time and time again — like dock doors and conveyors — and others that you might not ever have thought of — like smart shelves. Understanding the basic applications of the technology is important if you are going to plan for and design your RFID system effectively. You find out ways in which other people have applied RFID technology to their businesses so that you might be able to think of applications in your own environment.

Setting Up RFID Interrogation Zones

The RFID interrogation zone is where it all happens. It's the "bubble" created by a reader and antennas that allows a tag to be read and then that information to be automatically collected and sent back to an enterprise application like a warehouse management system or enterprise resource planning application. The interrogation zones are set up in fairly common areas that are generally *choke points:* that is, areas where all items have to flow through. A choke point can be something as big as a tollbooth on a highway or as small as a machine to dispense individual pills. Both of these applications use RFID for automatic data collection and counting.

Coming and going — Reading at a dock door

The dock door application of RFID is probably one of the best-known and most publicized uses of the technology. An *interrogation zone* (where tags are read) can be easily set around a dock door and, with very high accuracy, read a pallet tag. Figure 6-1 shows a common dock door setup. In many instances, with a properly tuned system, that same dock door portal can read every case on a pallet as it crosses the dock door.

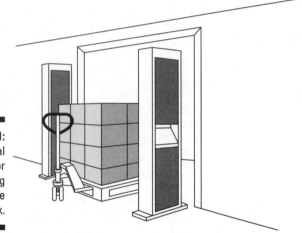

Figure 6-1:
A typical dock door setup using a premade RFID rack.

What you need for an ideal dock door setup

The dock door deployment works best with the following elements:

- ✔ **A high-powered UHF system** because it gives good long-range read capability for the cost of the system.

- ✔ **Ample power to the antennas in the interrogation zone** without being so strong that it crosses over into adjacent dock doors two and three bays down.

- ✔ **A reader set to constantly *poll*, or look for, tags** so that if something comes through the interrogation zone quickly, you maximize your chances of getting a response from the tag. It might take two or three tries to get a successful response, so you want to poll often enough to have at least a couple of attempts.

Setting up a dock door portal

You can deploy a dock door portal in many different ways. Here are the general steps that you follow:

1. **Decide on the area to be covered and the number of antennas to use.**

   Most of the UHF RFID portals used for dock door deployments have a total of four antennas, two on each side of the door. Sometimes they might have as many as eight antennas if the goal is to read every case on the pallet or perhaps to cover a larger area.

2. **Determine where to place the antennas.**

   Some dock doors require you to place an antenna overhead as well as on both sides to get adequate coverage and penetration into the cases. However, this setup is more problematic because every overhead antenna installation will be custom as opposed to using a prefabricated reader rack.

   If you can manage to keep portals just on the sides of the doors, many companies, such as Symbol Technologies (www.symbol.com), Venture Research, Inc. (www.ventureresearch.com), ODIN technologies (www.odintechnologies.com), and others make prefab rugged racks for the side of dock doors. These racks usually have an easy way of mounting antennas, protection to keep the hardware from being bumped into, a secure place to house the reader, and somewhere to mount lights or motion detectors. Having racks to set up and deploy in a very standard way can allow someone to set up readers and antennas in 10 or 15 minutes for each dock door instead of several hours.

3. **Install the antennas and make sure that they're *canted* (angled) slightly outward (pointing into the back of the truck) to help eliminate cross talk among readers at adjacent dock doors.**

4. **Perform a thorough path loss contour mapping (PLCM), as detailed in Chapter 7, to help create the correct configuration and setup and to make sure that the power and configuration settings are optimized to read everything at that dock door, and *only* that dock door.**

Although reading all the cases on a pallet is one of the biggest benefits from a business process perspective, this poses the biggest challenge in a dock door portal setup. To maximize your chances of reading everything on the pallet (instead of just the pallet tag), optimize the cases for the right tag and the right placement. Optimized cases improve your performance not only on individual items, but also on your full pallet amounts. Chapter 9 walks you through the specifics of finding the right tag, placement, and orientation for your products.

Your gateway to good reads — Other portals

The idea of a portal setup goes beyond just using readers at a dock door. Anywhere there is a choke point (or an area where everything must flow through) in a warehouse, building, highway, or process, there is a candidate for a portal. Some useful portal applications include

- **Doorway portals:** The doorway portal is quite useful for item-level tracking. A doorway portal can be deployed for everything from *asset management* (tracking the coming and going of property) to *security and personnel access* (making sure things that go out a doorway are supposed to go out a doorway). Very primitive forms of RFID have been used for decades in keeping store inventory safe. The tags used by companies like Sensormatic or Checkpoint in those applications are chipless and have only two bits of information: One represents store property and sets off the alarm, and the other represents purchased property and allows the person to walk through the doorway portal without setting off the alarm.

- **Security portals:** The Federal Government is always looking for ways to add security to Sensitive Compartmented Information Facilities (SCIFs). RFID can easily be a covert way to track laptops, hard drives, and even handheld devices as they enter or exit these secure facilities. Other facilities, like hospitals, track assets around the facilities by using active RFID tags. Hospitals use active RFID (or RFID-like) systems based on Wi-Fi because they already have many devices that operate in the UHF band. Most active systems operate at 433 MHz.

- **Luggage portals:** Figure 6-2 shows a luggage portal from the Hong Kong airport, a very successful use of Matrics/Symbol readers to track luggage. The setup of the portal is the same as a dock door portal with an overhead component. The antennas on all three sides contain a transmitting (Tx) and a receiving (Rx) element. The proper tuning of this system has enabled a nearly 100 percent read of all luggage through the portal. In fact, the tag designed specifically for this application has proven to be a great performer in many instances. See Chapter 5 for more on tuning and other details about how reader antennas work.

- **Car or bus portals:** If you live in a city with designated commuter lanes or secure access to major airports, they are most likely controlled by RFID systems. Buses on a tollroad or commuter lanes can often access special lanes by driving under a portal and having a gate automatically open after the RFID tag on the bus is read. Secure patrol cars in airports will sometimes have tags mounted underneath the car and be read by a reader in the pavement in front of a secure entrance.

Photo courtesy of Matrics/Symbol.

Figure 6-2:
One of the luggage portals at the Hong Kong airport.

The idea behind these portal applications is essentially the same as the dock door portal, but the configuration of the readers is usually slightly different because other portal applications are usually smaller than a ten-foot-wide dock door.

A common mistake is over-designing the system — namely, adding more antennas and extra readers. If you use more readers, tags, or power than you need, the different systems interfere with each other when you put portals side by side in a full RFID network. It takes only 100 microwatts or −10 dBm to power up a tag and get a successful response. So each portal needs to have that much power only in the areas where you want the tag to be read. To get the right amount of power at a dock door, for example, place two reader antennas on either side of the door, and make sure each antenna reads only just past halfway across the door. This is true for any instance where antennas are on both sides of the portal. Too much power is a bad thing, particularly if other portals are close by. Some of these engineering issues will be solved by Generation 2.0 readers, which have a feature for dense reader environments. You can expect to see these Gen 2.0 readers in late 2005.

Keep on rollin' — Setting up RFID at a conveyor

The conveyor is the second pitcher in the RFID rotation. The dock door might be the star player that is throwing the 95 mph fastball, but the conveyor has all the subtleties, complications, and effectiveness of a knuckleball from Red Sox pitcher Tim Wakefield.

The conveyor setup often consists of four antennas in a quad arrangement, as shown in Figure 6-3. This setup gives an RFID conveyor application huge advantages over a bar code conveyor solution because the RFID tag, unlike a bar code, does not have to be facing in a specific direction. You don't have to worry about properly orienting a package as it makes its way through a sort station or along a conveyor line.

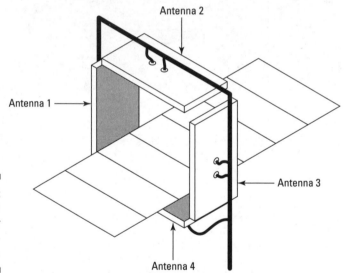

Figure 6-3: A basic conveyor interrogation setup.

Here are two important guidelines for positioning the antennas in a conveyor setup:

✔ **Set the antennas far enough away to have strong far-field communication.** This means that the antennas should be about 18 inches away from the edge of the conveyor. Even though the near field is about a foot (33 cm wavelength), metal and other objects can change the boundary (interfere with the far field communication) between the near field and the far field, so having a little wiggle room is a good idea. (For more on near- and far-field coverage, see "One at a time — Reading objects on a shelf," later in this chapter.)

✔ **Position an antenna underneath the conveyor to interrogate tags that may end up facing the ground.** Initially, I thought these antennas needed to be installed under a nonmetal roller of some sort, but I've found that they work pretty well underneath metal conveyor rollers. In fact, that is exactly how the largest retailers have them deployed in their distribution centers.

It's important to design your RFID network with the end in mind. Although you might set up only one or two read points for a pilot today, eventually you might have 100 read points in the warehouse or distribution center. When you build out your network, you will create an architecture that has critical dependencies and correlations. The conveyor is a primary example of using these relationships. If you have one reader in the same sort line or conveyor line as another reader, the reads on those two should be highly correlated. If they aren't, you need to address a performance issue. You want to be able to compare these two points in real time.

That's a wrap — Interrogating at a shrink-wrap station

From a business process standpoint, the difference between a negative return on investment (ROI) and a positive ROI might be the ability to read each individual case and track information at the case level. The best method for reading all the cases on a pallet is to set up an interrogation zone at a stretch- or shrink-wrap station.

The type of machine that wraps your pallets determines how you set up the interrogation zone. For example, if you have an arm that moves around the pallet, you might set up the antennas of the RFID reader on the moving arm. However, if you have a turntable that spins around a roll, you might fix an antenna off to one side of the machine.

The best way to set up an interrogation zone at one of these stations is with a combination of two antenna locations, as shown in Figure 6-4.

✔ **First location:** Affix an antenna to the arm that moves with the roll of shrink-wrap. This requires a little more time in the installation process but yields much better results than just setting up fixed antennas. It's critical to install the antenna cabling in as protected a manner as possible. To do this, you need to account for the movement of the arm by having extra cable that can follow up and down as you cable the antennas back to the reader.

✔ **Second location:** Set up the other antenna just next to the stretch-wrap machine. Make sure that the antennas are not in the way of the forklift as it drops off the pallet but close enough that the transmitting and receiving antennas can interrogate the pallet.

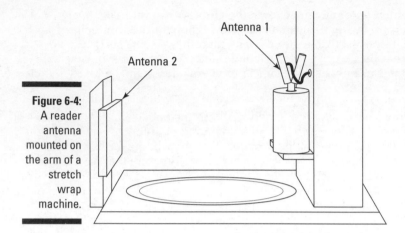

Figure 6-4:
A reader antenna mounted on the arm of a stretch wrap machine.

Antenna 1

Antenna 2

How the pallet is spun on the turntable and the fact that a receiving antenna moves close to the pallet (on the machine's arm) make for a very favorable setup to penetrate deeply into pallets full of cases in order to record accurate results. After the pallet is stretch-wrapped, many people apply an RFID pallet tag that summarizes all the contents of the pallet or indicates a standardized number (like an SSCC — serial shipping container code). This makes a great area to also set up a printer for additional tags and fits in well with the business process. A stretch-wrap station is a great place for a reader because the turntable of the stretch-wrap machine is constantly changing the location and orientation of the tags and giving the readers many attempts to read each of the tags as the pallet spins in the same spot.

One at a time — Reading objects on a shelf

The *smart shelf* was one of the first areas the scientists at MIT sought to conquer. Driven by early inspiration from Gillette, which has theft issues with razor blades because of their relatively high cost and small size (easy to pocket), the MIT team came up with several ways to read individual objects on a shelf.

Choosing the right frequency

High frequency (HF), at 13.56 MHz, works best in a shelf situation where it's important to know the location of the items but not to read across multiple shelves. However, HF can't cover as much area as ultrahigh frequency (UHF), 915 MHz. HF works best in the near field, approximately 1–6 inches from the antenna, and UHF communicates in the far field, about optimally a foot away from the antenna. HF is the appropriate choice for most shelf readers because it reads short range effectively and is less affected by metal and liquid. When you set up shelf readers, you need to make sure that the antennas are designed with the frequency, the distance of the tags, and the type of objects in mind.

Configuring a shelf reader

Reading tags at a dock door or a conveyor is like watching Jeff Gordon at the Daytona 500 — you've got to be looking every second, or you'll miss the car as it flies by. For a shelf reader, however, speed isn't an issue. Rather, *state change* (something changing from what it was before: on the shelf or not on the shelf) is the driving design factor. You don't need (or want) to interrogate the shelf's state hundreds of times every minute — doing so just creates tons of useless data.

You need to configure shelf readers to detect state changes or to poll at relatively long intervals. Some readers, like the Symbol AR-400, can poll continuously but report only when a tag comes in or out of the field. Other readers, like the Texas Instruments 13.56 MHz board, enable you to custom-program them to interrogate on various time settings. Choose whichever type of reader best fits your needs.

Setting up

If you have one antenna per shelf (as shown in Figure 6-5), you can usually use two or four antennas per reader, depending on the reader and frequency. To read the tags, the antennas typically cycle through a set order, such as the following: antenna 1 first, antenna 2 second, and so on. You can also customize the read cycle for your needs. For instance, if you have a top shelf in your store of DVDs and you want to track which ones customers pick up and look at, you might want to constantly scan that shelf. (The RFID crew at MeadWestvaco in Maryland has built some interesting shelf applications for DVDs using shelf antennas, if you are looking specifically for a shelf application.)

A HF tag works best when the magnetic field is perpendicular to the label (that is, they meet at right angles). If the magnetic field is parallel to the label, there is no coupling between tag and antenna and therefore no communication. This is why it is important to orient the tags at right angles to the radiation of the antenna.

HF tag designs are based on the theory of near-field communication and have antennas designed in a *multiturn planar coil,* or simply a multiple turn of the antenna material. An induced voltage powers the tag and enables it to communicate. This voltage is created through Faraday's Law of Electromagnetic Induction. At HF, where communication is in the near field, coupling volume theory is what dictates the design of the system, and the focus is on the energy stored per unit volume around the tag (the coil stores up energy). Conversely, at UHF, where communication is in the far field, radiating antenna theory drives the system design, and focus is on the electromagnetic power flow per area flowing past the tag.

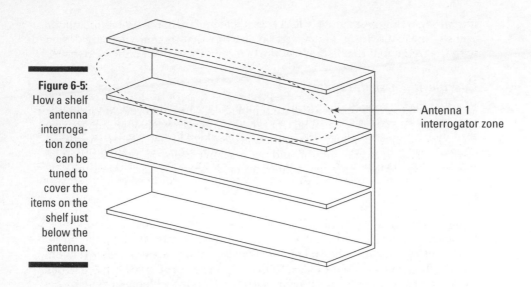

Figure 6-5:
How a shelf antenna interrogation zone can be tuned to cover the items on the shelf just below the antenna.

From Ski Resorts to Airlines: Applying RFID in the Real World

The ability to track and trace, follow and find, and sneak and peek are all enhanced by the use of RFID. Some privacy advocates worry that machines will magically attach RFID tags to your clothing or shoes as you walk through a store or around an office, but this is very unlikely. The limits of the RFID technology are well known and defined by the laws of physics; however, effective ways to use this technology are just beginning to emerge. The next few sections examine some real-world applications of RFID technology and how they benefit the companies involved.

Ski resorts

Ski resorts, hospitals, and water parks are all using RFID wristbands to follow patrons and mine information or eliminate payment steps. The benefits of RFID in this type of situation are twofold. RFID provides a convenience to the user and, at the same time, creates a more efficient operation for the business.

Think of using RFID at a ski resort. A family shows up and gets mom, dad, and the two kids wristbands with not only their lift tickets for the day attached to them but a certain amount of "mountain dollars" associated with the band as well. Each of the kids has $25 in case they want to get a soda or buy lip balm (but not both — if you've been to a ski resort lately, you know that would require $50). As it gets colder and the family throws on extra jackets, they don't have to fumble for their lift tickets because the RFID reader can penetrate easily through a ski jacket. And if one of the kids gets lost, the parents can go to the ski patrol and find out where and when Junior was last scanned, so they know where to start looking for him.

The benefits to the ski resort are that the speed of people onto and off of the lift is greatly increased, and lines are diminished. They also get a lot more information about each skier, such as what their patterns of behavior are, because they can track each individual and their preferences. The lines are shortened because RFID is not line of sight, and people don't have to fumble for their lift tickets to be verified visually. Counterfeiting lift tickets is also eliminated.

Law enforcement

The U.S. Department of Homeland Security and the law enforcement community in general are keenly interested in using RFID to track everything from evidence to drug shipments. Here are a few RFID applications in the pipeline:

- ✔ **Tracking imports:** The Department of Homeland Security is supporting an initiative to put active RFID tags on all containers coming into the United States. This would enable our Customs inspectors to greatly increase their reach by sealing and verifying containers when they are packed and then entering the country as "trusted" cargo.

- ✔ **Controlling access to secure areas:** Certain vehicles at airports have RFID tags hidden underneath them that allow secure access to restricted areas. RFID readers are embedded in the roadway and determine whether a vehicle is authorized for access.

Pharmaceuticals

One of the biggest areas of promise for the use of RFID is in the pharmaceutical world. Currently, theft, counterfeiting, and diversion of expensive prescription drugs are driving up costs. Here are some ways that RFID is helping drug companies combat these problems:

- ✔ **Theft and counterfeiting:** One way for manufacturers to curb theft and counterfeiting is to apply tamper-proof RFID labels to these drugs. Pharmacies and hospitals verify the validity of their stock against a secure database. If this initiative is widely adopted, along with other methods like blister packs and chain of custody, this could completely wipe out counterfeiting.

- ✔ **Diversion:** A bigger problem that RFID can solve for the drug companies is *diversion*. This phenomenon arises from tiered pricing structures that the government imposes on pharmaceutical companies and the distribution infrastructure in place.

Here's how diversion works. Say that Hoboken V.A. hospital orders 1,000 pills from its distributor, Soprano Distribution. Soprano then contacts its supplier and places an order for the V.A. hospital, which means the pricing is 40 percent less than the normal commercial pricing. But rather than place an order for 1,000 pills, Soprano ups the order to 2,000. Soprano ends up getting 2,000 pills at 40 percent off; it then sends 1,000 to the V.A. hospital and keeps 1,000 to sell to its commercial customers at a much higher profit margin.

Although this is an oversimplified example, it happens all the time. In fact, some of the drug manufacturers have even bought back their own drugs from distributors when they were short of product — for more than they sold them — because of this problem.

RFID and a centrally managed and secure database could solve this diversion issue by individually labeling each drug with its own identity information and then recording each time it goes to a particular distributor and what the original pricing was meant for. That way, if a drugstore scans in a new delivery of a drug that's incorrect, it can send an alarm or notification. The feds can then work back the chain of custody based on earlier scans to see where the diversion took place and, yet again, bust illegal distributors.

Additional business applications

Companies can also benefit from RFID in the following areas:

- **Hazardous materials and recalls:** A number of areas relating to hazardous materials and recalls hold huge promise for RFID:

  - A great example is the work the RFID team at Michelin has done to embed a passive RFID tag into every tire they manufacture. Although not yet in production, this has huge implications for maintenance, shipping, recalls, and safety inspections.

  - Companies manufacturing batteries that must meet disposal regulations can embed RFID tags in their batteries and provide an incentive for consumers to return them for credit on their next battery purchase. This allows the manufacturer to more accurately track the returns and disposals even after the outside bar codes and human-readable text have long worn off.

- **Warranty verification and returns:** The Computer Technology Industry Association (CompTIA) is working with manufacturers to use RFID on everything from work-in-process parts to warranty verification and returns. Imagine not having to go through a hassle at your local electronics store when you return the surround-sound system that went on the fritz. They scan the RFID tag on the receiver, note the date and time you purchased it (information that is kept on a store database), and give you a new one.

✔ **Manufacturing:** Many heavy manufacturing industries are seeing the benefit of RFID as well. Automotive manufacturers have used active RFID tags for years and now incorporate passive tags on car windows as they travel through the assembly process to track the entire work in process of the car from start to finish.

✔ **Maintenance:** The airline industry is finding great benefits in its maintenance process. You know those oxygen masks you hope never drop down when you're flying on a plane? They are attached to bottles of oxygen above the seat that have various expiration and refill dates. To check these bottles, airline personnel used to unscrew each plastic cover and visually inspect the date. It took hours to check a large plane. Now they put an RFID tag, which can be read through the plastic cover, on each bottle and walk down the aisle with a hand-held reader to gather the exact same information in only minutes.

As RFID technology becomes more pervasive (and it's possible for you to set up RFID systems without a degree in physics), RFID will find its way into our homes and offices in everyday applications, much like the Internet and wireless remote controls already have. The first companies to adopt and optimize the technology are the ones who will leverage the competitive advantage any disruptive technology can offer.

CASE STUDY

Using RFID to track meat

A meat-packing company approached my company to design a unique RFID application for tracking a meat product stored in a dumpster-like container as it moved from a cold-storage area into a processing area. The company wanted real-time assurance that the correct container was delivered to the correct processing machine in the correct sequence. The application needed to be operator-independent and totally automated.

In order to set up the RFID tracking system, I needed to devise a way to carefully tag the containers so that the tags could not potentially fall into the product and contaminate the system, and by association the meat. The tags also needed to withstand cleaning conditions, steam, and high pressure. Here are the details of the setup:

✔ The tag was a stick tag sold by Intermec, capable of being attached directly to metal.

✔ The native read distance was 3–4 meters.

✔ The tag was attached with silicone adhesive.

✔ The reader system was a portal at the sliding door to the processing room. The equipment was a single Intermec fixed reader with two antennas facing each other from opposite sides of the 12-foot opening. The reader was enclosed in a NEMA 4 box and hard-wired to house electrical components. The data management system was RS-232 direct to a local database.

The company also wanted me to attach tags directly to the meat hooks in the slaughterhouse. These tags needed to be hermetically

(continued)

sealed to avoid any contamination, and they also needed to be small. The goal was to transfer the ear tag information to the hook to follow the cow all the way through the plant. Here's the setup that I used:

- ✔ I used 2.45 GHz inlays from Intermec and molded them directly into plastic parts specifically designed for these particular hooks.

- ✔ The attachment was done by a machine, and the tag needed to also survive tremendous physical abuse.

- ✔ The readers were NEMA 4–encased and distributed in a number of places in the plant.

Another popular use of RFID technology is tagging reusable pallets and containers. My experience with tagging pallets for another company was no small feat. The company wanted to use a single tag for an entire pallet. I proposed two options: place the tag in a plastic end plate or place the tag in a channel in the wood. The company opted for placing the tag in a channel. The biggest challenge was tagging fresh pallets. A normal pallet is at least a year old and has a moisture content of 6–12 percent, but a fresh pallet can approach 30 percent moisture. This amount of moisture can cause up to 6 dB in tag or system performance, which is enough to limit read distance by several feet. I set up a system with a single reader to read the pallets on a conveyor. This setup yielded good results, but the moisture content of the new pallets was something I hadn't bargained for. This was one of those applications that helped drive a scientific process for testing products to be tagged, to avoid surprises like the moisture content.

Part III
Fitting an RFID Application into Your World

The 5th Wave   By Rich Tennant

"Can't I just give you riches or something?"

In this part . . .

If you feel like you are taking full strides when you start reading this section, by the time you finish reading it, you will feel like the RFID equivalent of the Red Sox's Johnny Damon, stealing second base with lightning speed. If you just want to get started setting up the system, this is the part you probably want to turn to first. This part covers the real nuts and bolts of how the technology works.

Part III offers you step-by-step instructions on how to assess your environment and choose the right tag and readers, and then helps you figure out the best middle-ware to connect all of this back to your key applications.

Chapter 7

Seeing the Invisible: The Site Assessment

*W*hen you think about installing an RFID system in your warehouse or distribution center (DC), you may think that the only obstacles you have to worry about are a fast-moving forklift or an angry dock worker. But the real villain is invisible, and it's called *ambient electromagnetic noise* (AEN). This rogue AEN consists of electrical and magnetic waves that certain electrical devices generate and propagate through the air. In a typical warehouse or store backroom — with conveyors, electric doors, sorting machines, infrared scanners, real-time location systems, alarm motion detectors, site radio communication, and a plethora of other systems generating electrical noise — it's easy to imagine how these electromagnetic waves fill the space around you.

Because these systems (which you already have in place) can create AEN with radio-frequency signals in the range that RFID communications use, you have a real potential for signal interference. This interference may adversely affect your new RFID network. And the opposite is also true: Your fancy new high-powered RFID system may adversely affect your existing systems.

To find out exactly what's happening with AEN in your environment, you need to test the location where you're deploying RFID. This testing is commonly called a *site assessment* or *site survey,* and it enables you to diagnose (and plan to avoid) the potential operational problems associated with installing an RFID system.

In this chapter, I show you how to perform a site assessment and thereby see the invisible AEN present in your environment. Specifically, I help you identify the potential sources of electromagnetic waves and decide where in your facility to test for AEN. I outline the process for setting up test equipment (like a spectrum analyzer), measuring AEN over time (by performing a Full Faraday Cycle Analysis), and evaluating the results. Then, based on your results, you find out how to perform RF path loss contour mapping, which helps you determine what antenna design, placement, and tuning you need. This leads to the optimal reader configuration required for a successful RFID system installation, and also gives you the information needed to select the right reader for your particular environment.

RFID isn't like a fast food uniform — one size doesn't fit all. There is no such thing as the perfect solution for every application, at least not yet. So you need to understand your environment and how your products react in that environment before you can choose the right tags and readers.

Planning for Your Site Assessment

Understanding the working environment is the starting point when designing and implementing a successful RFID network. The best way to do this is to perform a radio frequency (RF) site assessment to ensure that the proposed RFID installation can operate within the planned environment at optimal performance. Sounds easy, right? Unfortunately, the complexity of invisible electromagnetic waves in your working environment makes sure that it's not.

The future of RFID in the ISM band

RFID in the United States operates in the 902–928 MHz band. This is an unlicensed band that many other systems operate within. Because many systems operate in this band, the FCC requires that any device in use, according to regulation 15, uses *frequency hopping*. This means that a device cannot transmit on any one of the 124 channels for longer than a fraction of a second. The signals literally have to hop from channel to channel in a pseudo-random manner. Because of the explosion of popularity in RFID systems, the FCC is faced with some increasing challenges as the high-power RFID systems begin to compete with other devices for channels in the ISM band.

One of the protocols most talked about within the ISM band is a low-power wireless system called Zigbee, a next-generation Bluetooth type of technology. The problem is that Zigbee and other systems can't work when a high-powered RFID system is operating. So what is the solution?

The best solution would be for the FCC to allocate a small portion of the 902–928 MHz frequency specifically to RFID (say from 920–928 MHz). This would allow RFID to have its own bit of spectrum and not interfere with other unlicensed technology. This would also allow RFID to broadcast in narrow band (using one channel for extended periods of time) and avoid frequency hopping. This means that many systems could operate in the same warehouse without interfering with each other.

The site assessment is important because it enables you to see the invisible forms of radio frequency communication already present — intentionally or otherwise — where you want to install your reader network. The vast majority of RFID systems operate in the Industrial, Scientific, and Medical (ISM) band of 902–928 MHz (or 864–870 MHz in Europe), which is an unlicensed frequency in the United States. Because it's unlicensed, your RFID systems have to get along with other devices that use this same unlicensed band, including cordless phones, long-range radios, barcode devices, alarm systems, real-time location systems, and many other wireless gadgets.

With all these devices vying for the same communication space, intermodulation (that is, communications competing for the same channels — like when you were on an old analog cell phone and you could hear someone else's conversation) and data collision are inevitable if more than one system is operating in that unlicensed ISM band. Data collision is messy, costly, and a real bear to clean up. But you can prevent a nasty clean-up job by performing a thorough RF site assessment. Consider the following points to plan for and execute a successful site assessment:

- ✔ **Go to your business-process map (outlined in Chapter 17) where you determined the target locations (known as *interrogation zones*) for installing your RFID systems.** These locations vary from warehouse to warehouse and need to have a power source and Ethernet connectivity for the readers (unless you have readers with wireless connectivity). Commonly, RFID interrogation zones are placed at dock doors, shrink-wrap stations, conveyor lines, and inventory shelves. After you identify the target locations for RFID in your facility, you carry out the RF testing procedure at each one.

- ✔ **During your site assessment, look for two things: the strength of the waves that propagate through your potential interrogation zone and the frequency those waves broadcast over.** Plan to perform the RF assessment at each target location one at a time so you can get a good picture of the relative strength of signals in each area. The higher the prevalence and strength of waves in a particular band, the more difficulty you'll have implementing a successful RFID network.

- ✔ **Look at AEN over a period of time, during all your business operations.** Normally this is over the course of 24 to 48 hours. When doing a site assessment, many people make one big mistake: They simply take a snapshot of the AEN at a particular point in time (which is like choosing golf clubs to use for an entire course by looking around the seventh green). Instead, you should follow a process that takes a full look at the entire electromagnetic cycle as your facility goes through its normal business operations (which is more like walking all 18 holes of the course and taking notes before you play it).

This method of capturing all the relevant data is called a *Full Faraday Cycle Analysis* (FFCA). This fancy-sounding name represents a way of gathering time-dependent spectrum analysis data across a specific band of operation at the exact locations where you'll be setting up an RFID interrogation zone. For your site assessment, this band is the 902 MHz to 928 MHz (ISM) frequency band. A Full Faraday Cycle Analysis ensures that, before you start building your RFID network, you'll have all the data you need to make the right decisions regarding the type of hardware you need and the way it should be configured. The analysis also helps you understand any challenges you might face or systems you'll have to work around.

The FFCA also must take into account the entire space of your operating environment. The best way to do this is to begin looking at AEN at the four corners of a warehouse and moving toward the center, taking additional measurements and recording the relative strength. It is important that you measure the initial four corners over all work shifts. I've seen airport radar repositioned at a certain time each day, shipping companies that come in with handhelds, and other sources that can interfere with an RFID system.

Getting the right test equipment

The good news about an RFID deployment is the same as the bad news: The frequency band used requires no operating license or permissions from any governing body. Although the FCC has strict rules governing operation in this FCC-allocated band, no operating license is required to use the 902–928 MHz range. (To see the FCC rules, go to http://wireless.fcc.gov, click Rules and Regulations on the left, and scroll down to Part 18 — Industrial, Scientific, and Medical Equipment.) So before you deploy an RFID system, you need to become an RF detective by setting up test equipment to find out what other signals are active in the target area and might affect RFID performance.

You need the following equipment, shown in Figure 7-1, to correctly set up an RF site assessment:

- **Spectrum analyzer (SA):** A device that measures the relative strength and specific bandwidth of communication across a given range (in this case 902–928 MHz) and that serves as the data-logging mechanism in the testing setup for your site assessment.

- **¼-wave or ½-wave dipole 915 MHz antenna and ground plane plate:** The antenna is attached to the center of a ground plate (as shown in Figure 7-1) to properly load the antenna. (The ground plane doesn't need to be perforated, like the one shown in Figure 7-1.) The antenna is also attached to the spectrum analyzer by a coax cable. The antenna listens in 360 degrees to all the ambient electromagnetic signals, and then sends those signals back to the spectrum analyzer for display.

If you are using a ½-wave antenna, you don't need a ground plane. You can attach the antenna directly to the tripod. With the ¼-wave antenna, you need to use a plate of metal at least 33 cm by 33 cm as the ground plate.

✔ **Tripod stand:** The mechanism that supports the antenna in the center of the target location. A tripod should be sturdy enough to hold up the antenna and flexible enough to go from a foot or two off the ground up to five or six feet high. A high-quality camera tripod usually does the trick.

✔ **Laptop computer:** The laptop is used to log time-based data captured by your RF testing setup. The computer is usually connected to the spectrum analyzer by an RS-232 or Ethernet cable. If you use an older — usually cheaper — spectrum analyzer that doesn't have the option of connecting directly to a laptop, you can do without the laptop and use a digital camera to take pictures of the SA's screen at various intervals to record the data. Although this is not as elegant as having the laptop create the time-based graphs, it is equally effective.

RF testing equipment has been around for decades and is widely used for everything from ham radio design to Wi-Fi network setups. There are a number of sites online where you can find new equipment, but the best deals are usually found on eBay, which is a fine place to get good equipment inexpensively. If you go this route, search for a spectrum analyzer that covers the 902–928 MHz band and make sure that it has been recently calibrated. If you want to buy this equipment new, National Instruments, HP/Agilent, and Tectronics all offer a good variety of choices.

Figure 7-1:
To test for AEN, use a spectrum analyzer and monitor (A) and a ½ wavelength dipole antenna mounted to a tripod and grounding plane (B).

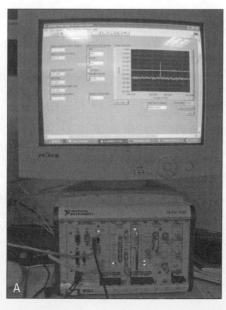

Hopping around the frequency band

You may hear people say they are setting up an ultrahigh frequency (UHF) RFID system at the 915 MHz frequency. This statement is only partially true. Because the FCC allows many unlicensed devices to operate at that frequency, it requires that no single device broadcast for more than a certain length of time. If only one frequency, or *channel,* was available, and a device couldn't broadcast for a more than a split second, it just wouldn't be possible to put many devices on that channel. That is why devices are designed to broadcast across a range of channels.

The process of moving through each channel is called *frequency hopping,* and most devices in this band stay on one channel for only 200 milliseconds or so. So to catch all the broadcasts, you need to measure not just the 915 MHz channel, but the whole ISM range — from 902 MHz to 928 MHz. (That's 13 MHz on either side of the center frequency, or a 26 MHz span.) The span you need to measure is more than twice that (a 60 MHz span), however, because other licensed frequencies may be operating close enough to your ISM band and with enough power to interrupt communications.

Setting up for RF testing

As you set up your RF testing equipment, keep in mind that you want to gather data over a period of time representative of the normal business cycle. Make sure that normal (or close to it) operations can take place after you put your testing equipment in place.

Although it might be easiest to test a warehouse on a Saturday morning when no one is around to get in your way, that is the worst possible time to get a true picture of the RF noise that is likely to occur. As crazy as it may sound, test during the busiest time for your location, and the data you gather will make the setup and deployment of your network easier by an order of magnitude.

Follow these steps to set up the test equipment:

1. **Place the ¼-wave dipole antenna on the ground plane plate and attach both to the tripod (or if you're using a ½-wave antenna, attach it directly to the tripod) so that the center of the antenna is in the center of the target area, as shown in Figure 7-2.**

   The target area is where you would like the RFID tags to be read — usually 3 feet above the ground for a dock door, 12 inches above a conveyor, and so on. Because you want the antenna located as close as possible to the center of the interrogation zone, the best way to mount the antenna is usually with the tripod contorted in one way or another, so make sure that you get a tripod that's easily adjustable. You also may have to get creative and hang the antenna from above to get it in the middle of the interrogation zone.

Do not hang an antenna from its coax cable because doing so may interfere with the signal and communication and not accurately ground the antenna. Instead, use a piece of rope or other nonconductive material to hang the antenna.

Figure 7-2:
An antenna at the center of a target area.

2. **Connect the antenna and ground plane to the spectrum analyzer by screwing the coax cable into the input port on the spectrum analyzer.**

As with all RF equipment, never connect or disconnect an antenna to or from a powered-up device, even if the device has fault protection. Although most of today's electronics have solid protection, connecting an antenna with the power off is a good habit to get into that may save you a few thousand dollars' worth of fried equipment.

3. **Attach the laptop computer to the spectrum analyzer, using either an RS-232 connector or Ethernet cable, and power it up.**

4. **Power up the spectrum analyzer and tune it to a center frequency of 915 MHz.**

See the sidebar, "Hopping around the frequency band" for the hows and whys of tuning to this frequency.

5. **Finish setting up the spectrum analyzer by doing the following:**

   • Set the span to 60 MHz. This setting ensures that the analyzer duly notes any AEN on either side of the 915 MHz center frequency. You want to know if any signals are even close because a device broadcasting at 901 MHz can cause interference.

   • Set the resolution bandwidth to 100 kHz to ensure that you record reasonable levels of interference.

   • Set the video bandwidth to 30 kHz to obtain a smooth plot on the spectrum analyzer.

- Set the amplitude attenuation to 0 dB so that the spectrum analyzer displays a discernible signal-to-noise level (making it easier to see the interfering signals above the noise floor).

- Turn on maximum hold so that you can capture the energy of every channel in the band you are testing.

After the antenna is in the middle of the proposed interrogation zone and attached to your correctly tuned spectrum analyzer, you should see the virtual screen on your laptop, or the video screen on the spectrum analyzer should be active. If you see either one of these, you're ready to start the testing.

Measuring for AEN during Normal Operations (And Beyond)

The Full Faraday Cycle Analysis is the foundation for building an RFID network, and the goal of that analysis is to have a perfect foundation for the structures being put on top of it. For a full analysis, you need to do the following:

- Identify all the ambient electromagnetic noise within the facility.

- Log data over the course of a full business cycle (all the shifts) to understand any changes that happen at different times of the day.

- Measure specifically at each interrogation zone to correctly assess the strength of signals relative to where you'll install readers.

- Rove around the facility to make sure you looked in every nook and cranny for rogue AEN.

- Triangulate any sources of interference while roving about the facility (in other words, get closer and closer until you find the source of the interfering AEN).

- Run all the possible machinery and equipment that is likely to make electronic noise in the interrogation zone while recording the data to make sure no potential source is overlooked.

- Address the potential interference found by either eliminating it (for example, upgrading a wireless barcode system from 915 MHz to 2.4 GHz or finding a creative workaround (such as deploying only hand-held RFID readers that do not broadcast at the full power of a fixed location reader and don't create as much interference).

- Map out the interrogation zones on computer-aided drafting (CAD) drawings or blueprints and make sure that those zones are noise-free.

With your equipment set up, you're ready to begin. The following sections explain in more detail how you accomplish each part of the analysis.

Testing key points around the warehouse

As you test, you need to log the data:

- ✔ **If you are using a laptop to log the data or employing a software-based spectrum analyzer (like a National Instruments model),** you should set your virtual monitor to record information every hour and actively log it to the hard drive.

- ✔ **If you are using a spectrum analyzer without a laptop,** you should come back and take a digital picture every two hours over the course of the normal business cycle. After taking the picture, clear the video display by resetting the video screen and begin collecting data again.

This initial measurement process is effective, but only represents one data point in a facility. How large the warehouse is and how strong interfering signals are will determine what you pick up from that one location. So if you have a large warehouse, you have to set up the same test procedure at several locations within the warehouse to increase the accuracy.

You want to make sure you test for AEN close to all the potential RFID reader interrogation zones (usually the dock doors, conveyor or sort stations, and so on).

If you do notice any significant spikes on the monitor, you've caught some interference. It is important to note the location, the time, and the frequency of the interference and try to map a pattern (for instance, does the spike occur every hour when the security guard makes his rounds?) or try to narrow the time period in the next day or two to figure out the source of interference (did a FedEx truck arrive to pick up packages during the window when you had interference?).

One drawback to this static testing methodology is that it is difficult to find the location of any interference. That's why you follow up with other tests, which I explain in the next two sections.

I've been a wild rover for many's a year

The next step toward increasing the accuracy of your Full Faraday Cycle Analysis is to take a roving data capture of AEN. This test is particularly effective if you share a facility with other tenants who may be running systems that are separated from yours by only sheetrock walls. To take a roving capture, you need the following:

- ✔ A portable uninterruptible power supply (UPS) or similar battery backup device. You need the ability to power the spectrum analyzer for 20–30 minutes. These are available from American Power Conversion for under $200 at most computer stores.

- ✔ A golf cart, shopping cart, or similar means of wheeling your equipment around a warehouse safely.

- ✔ A willing friend to act as a human tripod and carry the antenna around next to the golf cart.

- ✔ Blueprints or CAD drawings of the facility come in handy, too.

To do the roving capture, follow these steps:

1. **Set up the spectrum analyzer in the exact same manner as you did for the static capture, except the power source and the antenna are both mobile.**

2. **Rove around the facility, watching for noise in the spectrum analyzer's display.**

   Because you are moving about the facility, you need to watch the display closely for noise in the ISM band and notice where that noise occurred.

3. **If you do find a source of interference, make sure you note the exact location on your CAD drawing or blueprint and the strength in –dB or decibels (which is how AEN is measured, just like sound waves are measured in –dB).**

4. **Begin moving away from that original spot of interference in 10- to 15-foot increments in four directions. In other words, move north 10 feet, go back to the original spot, move south 10 feet, go back . . . you get the idea.**

   This gives you four read points equidistant from the original interference location to compare relative strength. If any one of the four points is stronger than the first in terms of AEN signal strength, the interference is coming from that direction. You need to repeat the same process from the new location (the one with the stronger signal) in order to triangulate the source of the interference; in other words, keep moving toward stronger signals until you find the source.

It helps if you have blueprints or CAD drawings of the facility. If you encounter any random noise, you can mark the exact location where you picked up the interference and the exact strength of that location. Doing this in multiple locations will also help you pinpoint the source of interference based on signal strength. Having blueprints will also help you lay out your design and location of the interrogation zones. If you don't have CAD drawings or actual blueprints, a simple diagram drawn to scale will help in the planning stages.

I don't hear anything; time to make my own noise

Now comes the fun part: making some noise. Mobile mechanical devices (like forklifts, hand trucks, and the like) may pollute your potential interrogation zones. To find out exactly what the effects of these devices are, you need to bring those devices into the zone, run them full power, and see what sort of AEN they create. Occasionally, the testing done during the normal business cycle might leave out a business process, like running a floor cleaner that's used only one day a week, so now is the time to make sure you cover every possible source of interference

You need the spectrum analyzer, antenna, and laptop computer used in the earlier stage. To do the test, follow these steps:

1. **Set up the spectrum analyzer in your interrogation zone by following the same steps I outline in "Setting up for RF testing," earlier in this chapter.**

   All the settings for MHz, resolution bandwidth, video bandwidth, and so on are the same.

2. **Make some noise. Whatever business machines you've decided to use, now is the time to fire them up and move them through and around the interrogation zone.**

   This includes driving in and out of dock doors, raising and lowering forklift blades, turning around, and so on.

3. **As you operate the machines in the interrogation zone, make sure you record the results. After a few passes through the area with various machines, stop recording the data.**

4. **Any data that shows up on the spectrum analyzer but didn't show up during the normal testing of the same area suggests that some form of interference is present. The next section explains how you ensure it doesn't become a problem.**

Solving interference problems

If at any time during the FFCA you discover interference, you have two choices:

- ✔ **Determine the cause of the interference and remove it.** If a forklift, an electronic scale, or an industrial magnet is causing issues, you may need to make sure it doesn't operate in the interrogation zone. Isolating the guilty device is an easy way to fix the problem. Sometimes this is as easy as moving it a few yards away from the interrogation zone; sometimes it may need to be moved clear across the building.

- ✔ **Work around it.** Working around it is the more difficult choice. Workarounds are difficult because a warehouse RFID system is designed to get reads over open distances of several feet or more. Shielding entire dock doors or conveyor stations from AEN is usually impractical. You might be able to find the source of the noise (such as an alarm system or hand-held scanners) and shield only those interferers. A common problem is that overnight carriers' equipment is very close to the 902 MHz band, and if you do a lot of parcel shipping, you may need a dedicated location that is not near an interrogation zone for these carriers to use.

You may find that wireless barcode systems, which are common in warehouses, are operating over the UHF unlicensed band as well. If this is the case and you're planning a full RFID deployment, you can pretty much plan on calling your bar code vendor and asking for an upgrade to 2.45 GHz. Many early adopters of RFID had to go through this very exercise after they realized what was messing up their bar code readers. If you don't upgrade, hand-held bar code scanners continue to transmit across the ISM, but the higher power RFID readers will use more channels. Eventually, as more RFID readers are deployed, the RFID readers will totally drown out the bar code readers.

Testing to Plan Your RFID Installation

With the Full Faraday Cycle Analysis complete (see the preceding sections for details), you have the blueprints for building an RFID network. You understand what you have to deal with. Now you want to use the data you gathered to design the interrogation zones (the areas where you'll set up RFID readers to read tagged items), which is like creating a perfect set of footings for the structures being put on top of your foundation. (For you artistic types, after the site assessment is done, it's like starting with a fresh canvas for your electromagnetic work of art.)

Designing the interrogation zones requires two steps:

1. **You begin by propagating a perfect RF wave around the area in which you propose to set up your readers.**

2. **Then you create a contour map that helps you determine how those waves are going to behave in that specific location.**

Multi-pathing is a phenomenon that occurs in RF communications when waves of matching frequency collide and drown each other out, causing what's called *nulls* or *null spots*. The RF path loss contour mapping helps you determine where those spots are likely to be.

In the following sections, I explain these steps in more detail.

Gathering your equipment

It takes more than breadcrumbs to map the path of an RF signal. Propagating the perfect RF wave and creating the contour map requires tools, and lots of them. Here's a quick list of items you need:

- **Spectrum analyzer:** A device that measures the relative strength and specific bandwidth of communication across a given range (in this case, 902–928 MHz) and captures data as you test.

- **Signal generator:** A specialized device that produces RF signals at preset frequencies, strengths, and durations. The signal generator is hooked up to a ¼-wave dipole antenna via a coax cable and will be used to transmit the generated RF field.

- **Circularly polarized UHF antenna:** The antenna used by any UHF reader, or one ordered directly from a company like Cushcraft or Sensormatic, is attached to the spectrum analyzer to measure the RF field received from the signal generator.

- **¼-wave dipole 915 MHz antenna and ground plane plate:** Again, you attach the antenna to the center of the ground plane plate and then attach the antenna (mounted on the ground plane) to the signal generator. The antenna radiates an RF field in 360 degrees.

- **Two tripod stands:** One stand supports the antenna in the center of the target location, and the other stand supports the UHF antenna at the outside of the interrogation zone. A high-quality camera tripod is often sturdy enough to hold up the antenna and flexible enough to go from a foot or two off the ground up to five or six feet high.

✓ **Laptop computer:** The laptop is used to measure the relative strength of the signal produced by the signal generator. If you decided to get an older spectrum analyzer that doesn't connect directly to a laptop, you can use a digital camera to take pictures of the SA's screen at various intervals to record the data.

At the end of the testing, you also use the laptop to take the data produced by the test and map it in a spreadsheet program like Excel or Lotus.

Comparing the perfect signal to the actual signal

After you have performed your Full Faraday Cycle Analysis and you are sure that no ambient noise is polluting the interrogation zone, you can start mapping out what the interrogation zone will look like. This processes is called *RF path loss contour mapping,* which basically involves mapping out the RF path where it varies from that perfect RF field.

The RF field propagating from an antenna is shaped like a giant pear, with the stem attached to the antenna and the fat part heading off into space. The items with RFID tags on them will move through the fat part. This RF propagation bubble's size and shape depends on the type of antenna you use, and it changes dramatically when there is anything within the bubble to reflect or absorb the RF waves. Figure 7-3 shows a graphical representation of what the RFID interrogation zone looks like.

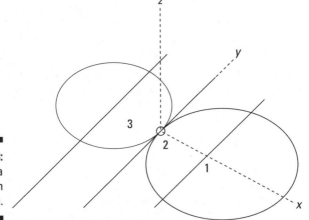

Figure 7-3:
Antenna
propagation
pattern.

RF path loss contour mapping enables you to understand how things around a proposed interrogation zone distort that perfect shape. You use this information to determine exactly where to place antennas and how much power you need to create a signal.

Setting up the equipment

Here's how to set up your equipment so you can start testing:

1. **Place the ¼-wave dipole antenna on the ground plane plate and attach both to the tripod stand so that the center of the antenna is in the center of the target area, as previously illustrated in Figure 7-2.**

   The target area is where you would like the RFID tags to be read — usually 3 feet above the ground for a dock door, 12 inches above a conveyor, and so on. Because you want the antenna as close as possible to the center of the interrogation zone, the best way to mount the antenna is usually with the tripod contorted in one way or another, so make sure that you get a tripod that is easily adjustable. You also might have to get creative and hang the antenna from above to get it in the middle of the interrogation zone.

2. **Connect the ¼-wave dipole antenna to the signal generator via the coax cable.**

3. **Set the signal generator to a signal of at least –14 dBm so that the UHF antenna can pick up a reasonably good signal.**

4. **Attach the laptop computer to the spectrum analyzer by using either an RS-232 connector or Ethernet cable. Power up the laptop.**

5. **Attach the UHF antenna to the spectrum analyzer. Power up the spectrum analyzer and tune it to a center frequency of 915 MHz and a span of 60 MHz.**

6. **Finish setting up the spectrum analyzer by doing the following:**

   • Set the span to 60 MHz.

   • Set the resolution bandwidth to 100 kHz.

   • Set the video bandwidth to 30 kHz.

   • Set the amplitude attenuation to 0 dB.

   • Turn off maximum hold to capture the energy received from the radiating antenna.

   See the earlier section, "Setting up for RF testing" for details on what each of these settings does.

Conducting the test

After setting up your equipment, you're ready to test. To understand how the test works, remember that the equipment is set up to simulate what will become your interrogation zones:

- ✔ The ¼-wave antenna attached to the signal generator pretends to be a tag bouncing back a mock signal that is read by the UHF antenna.

- ✔ The UHF antenna simulates the tag reader that collects information from the tags as the tagged boxes cross this zone.

To figure out the RF path loss, it helps to think of the interrogation zone area you want to set up as being roughly the shape of a pie and to think of the tag location as being at the center of a pie. Then divide the pie into eight slices (the slices would be cut at 0, 45, 90, 135, 180, 225, 270, and 315 degrees). Figure 7-4 shows where the ¼-wave antenna should be positioned in the center of the interrogation zone, and the eight spots around it show the location of the UHF antenna attached to the signal generator. Those eight corners are where you want to test the reaction of a 915 MHz propagation, or how well the RF wave travels back to be heard by the reader's antennas.

Antenna at a position around the pie

Center of pie

Figure 7-4:
Signal generator and UHF antenna setup diagram.

For the purposes of this first test, assume that you are trying to find out how a dock door and all the equipment around it will affect the placement of antennas, which direction they will have to face, and how much power each antenna will need. The location could be any potential interrogation zone, from a conveyor belt to a shrink-wrap station.

Follow these steps to conduct the test:

1. **Place the ¼ dipole antenna, which is connected to the signal generator, in the center of the proposed interrogation zone.**

2. **Connect the UHF flat panel directional antenna to the spectrum analyzer and mount it at the same height as the ¼-wave dipole antenna. Place it parallel to the dock door in the center (0 degrees in our Figure 7-4 pie).**

3. **Turn the signal generator on first to 902 MHz and record the results on the spectrum analyzer. Keep the UHF antenna in the same location and set the signal generator to 915 MHz and record the results. Lastly, repeat the process with the signal generator set to 928 MHz.**

   It is important to record the strength of the signal three times, once at each of the different frequencies.

4. **Relocate the directional antenna to each of the eight positions in the Figure 7-4 pie around the ¼-wave dipole antenna and record the signal strength at each position.**

   Keep the distance from the ¼-wave dipole the same. You want the distance to be the maximum distance you want to read from, which is usually half the total width of the dock door. Keep the directional antenna always facing directly toward the ¼-wave dipole.

   This will result in an eight-position contour map of the RF field strength.

5. **Put these 24 values into a spreadsheet program, such as Microsoft Excel, in the following format:**

Position in Degrees	Signal Strength		
	902 MHz	915 MHz	928 MHz
0	35	43	43
45	35	37	24
90	37	31	37
135	31	40	31
180	40	35	40
225	35	42	35
270	42	32	42
315	32	34	42

6. **Using the radar graph option in the spreadsheet program, convert the 24 results from your 8-position test into a radar graph.**

   The result should look something like the graph on the right in Figure 7-5. The graph on the left is a perfect RF field in a vacuum. The information you are interested in is how these two figures are different.

Putting your results to use

The Full Faraday Cycle Analysis and the RF path loss contour map are also the perfect blueprint for setting up other aspects of an optimal RFID network architecture. With those two tests completed, you can move on to the next key steps in deploying your RFID network:

  ✔ You can choose the best reader to fit your needs.

  ✔ You can set up the optimal configuration of those readers.

  ✔ You can verify and test the readers once they are set up.

The RF path loss contour map is an important tool for designing your reader interrogation zone. The ideal zone should be an equal bubble around the center pole (the left graph in Figure 7-5).

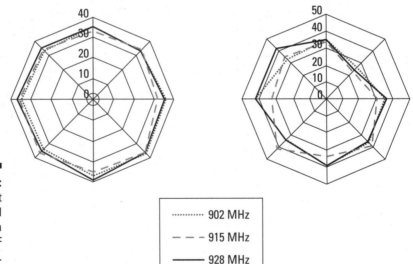

Figure 7-5:
The perfect
RF field
graph and a
typical RF
test graph.

············ 902 MHz

— — - 915 MHz

——— 928 MHz

You now need to compensate for any areas within that bubble that do not have equally powerful signal strength. If an area where you decide to set up an antenna is particularly weak, that will be a difficult area for tags to receive enough energy to power back a signal.

A passive RFID tag requires about 100 microwatts of power (or –10 dBm) to generate enough power and backscatter a signal to the receiving antenna of a reader. If you want to avoid problems of reading across multiple interrogation zones (like from one dock door to another), you need to make sure your power levels are below the –10 dBm level. This can be done by adjusting the power or shaping the field with antenna choice and direction or with shielding between the dock doors. Chapter 9 explains the specifics of reader setup and testing.

For example, if you refer to the graph on the right in Figure 7-5, you see that the signal reaching the point that represents 225 degrees is much weaker (or closer to the center) than the other points in the chart. The signal is weaker because something is either absorbing or deflecting the signal away from this area.

It is up to the reader configuration to compensate for this loss. To counteract that loss, the antenna located on that side of the dock door needs additional power compared to the other dock doors.

Chapter 8

Testing One, Two, Three: Developing Your Own Lab

*I*magine yourself in the woods of Switzerland back in the year 1307. You've got to shoot an apple off your son's noggin' or die. Do you just go out to your local sporting goods store, pick up a bow, and take a shot? Of course not, and neither would William Tell. He spent hours honing his skills and creating and testing bow designs before he became the expert marksman who let that famous shot fly. How do you hone your skills and create perfect designs for RFID before you make a huge investment in your production system? By putting together a well-planned and -organized lab. That is your key to becoming the William Tell of the RFID world.

As I mention throughout this book, understanding physics holds the key to a successful RFID deployment. But physics can be a fickle thing: Because you can't see how radio frequency waves change and behave, you have to find other means for knowing what happens with certain combinations of hardware, antennas, tags, and products. The best way to glean useful information is by having a consistent environment in which you can execute a repeatable process. A repeatable and consistent testing methodology allows you to change one variable at a time and compare results to understand the physics behind RFID.

Not only does a quality lab help you to choose equipment, design, and processes, but it is also a great place for your RFID team to discover new technologies and try new equipment in a low-risk, nonproduction environment. You save money by testing equipment and identifying what will work best for your RFID network before making a large financial investment in that equipment, and you can separate marketing hype from reality. The lab can also pay for itself quickly if you have a lot of items to test for tag placement and tag type — also known as *SKU testing*. (*SKU* stands for *stock-keeping unit*.) And lastly, a lab is also a great tool for getting the CEO excited about RFID.

This chapter takes you through the five steps of setting up a world-class lab and gives you some examples of test procedures you can use to compare equipment before making a buying decision. I tell you what equipment you must have, what is nice to have, and what is icing on the cake. In addition, I describe some of the test procedures we've refined at the ODIN technologies labs and explain how you can apply your lab results and knowledge to the real world.

To Lab or Not to Lab

Stop — before you spend any more time reading this chapter, you need to understand the three options for testing products, evaluating equipment, and trying different RFID configurations:

- ✔ **Use a third-party lab.** Third-party labs are great if you're testing just a few products or want the latest information on readers, antennas, and tags. As I write this book, three principal labs do scientific testing for RFID and also test for hire:

  - Met Labs (`www.metlabs.com`)

  - The University of Kansas, anchoring the consortium called the RFID Alliance Lab (`www.rfidalliancelab.org`)

  - ODIN technologies lab, which has been doing RFID the longest and produced the first head-to-head comparison of RFID readers and tags (`www.odintechnologies.com`)

- ✔ **Build your own in-house RFID lab.** The benefit of having your own lab is that you can maintain control over the testing, especially if you are concerned with competitors seeing a preview of new products or packaging. You also build a great amount of internal knowledge around the technology. The drawback is the initial expense and the operating cost to keep a lab running, train people, and recruit good talent.

- ✔ **Use hybrid approach.** In this case, a qualified consultant or existing lab sets up a lab for you. You use their testing software, services, and protocols to get the benefit of their intellectual capital but still maintain control and build knowledge.

Beyond a Swanky White Lab Coat: The Tools You Need for Successful Testing

Before setting up a lab, you need to think about what tools you need for a successful lab environment. If you're setting up a lab on your own, you need some of this equipment in order to build the lab. If you're using an outside firm to help you set up a lab, ask which of the following devices they plan on including. So get out your checkbook and let's go shopping. First, here are the must-haves:

- ✔ An assortment of RFID readers that have
  - The ability to read Class 0, Class 1, and Generation 2.0 tags
  - The ability to read multiple protocols
  - Integrated printers
- ✔ An assortment of linear and circular polarized antennas — both directional and omnidirectional (many will come with the readers, others can be ordered directly from Cushcraft or Sensormatic)
- ✔ A mixture of RFID tags
- ✔ An application server and database server
- ✔ Several hubs — USB, RS-232, and RS-485
- ✔ Several serial-to-Ethernet converters
- ✔ An assortment of zip-ties and plastic bands
- ✔ 1-inch thick colored tape
- ✔ 25 feet of 2.5-inch PVC pipe
- ✔ Ten 2.5-inch 90-degree PVC joints
- ✔ Four 2.5-inch PVC T-joints
- ✔ Four camera tripods

If you're aren't yet familiar with the different readers, antennas, and tags, flip to Chapter 5, where I explain these in more detail.

The following list shows the tools that are nice to have, but that are not entirely necessary. (I explain these in more detail in Chapter 7.)

- ✔ Lab-grade spectrum analyzer that can operate up to at least 1.5 GHz
- ✔ Lab-grade signal generator that can operate up to at least 1.5 GHz
- ✔ RF power meter with power head for frequencies up to 1.5 GHz and up to 5 watts of power

- ✔ RFID racks

- ✔ Cyclotron or similar instrument to accelerate the products with variable speed controller (Chapter 10 talks about ODIN technologies' Cyclotron and the benefits of using a Cyclotron instead of a loop conveyor.)

- ✔ RF attenuator (This device is attached to the cable between the antenna and the reader and enables you to precisely control the power output.)

- ✔ Loop conveyor with variable speed motors and safety rails

Now, if you've got a great boss and a hefty budget, the following tools are the icing on the cake:

- ✔ Lab-grade oscilloscope with 150 MHz bandwidth minimum

- ✔ RFID testing software

- ✔ Multimeter

- ✔ Forklift

- ✔ Server rack or cabinet

- ✔ Computers, routers, and spare monitors

- ✔ Anechoic chamber (see the sidebar, "Building an anechoic chamber")

- ✔ Your own dedicated lab rat

Combining the must-haves and nice-to-haves with proper testing methodology and procedures gives you a lab that'll stand up to any lab in the commercial world. Add some innovative testing software, and you can produce results in-house that others in the commercial world pay hundreds of thousands of dollars to get.

Setting Up Your Lab

After you gather the tools on your shopping list, you're ready to set up your lab. In the sections that follow, I walk you through the five steps of setting up your world-class lab:

1. Find the perfect location.

2. Design the physical layout.

3. Set up the test equipment.

4. Build specific test equipment.

5. Develop and implement standardized test procedures.

The steps are laid out in a specific order to make setting up your lab as easy as possible.

X-ray marks the spot: Find the perfect location

The first step in setting up a lab is deciding where to put it. Most companies put a lab in either a warehouse or part of an office building. Both locations have their own challenges.

Examining the ideal conditions

Before weighing the pros and cons of setting up a lab in a warehouse or office building, it's important to consider the common attributes for a good lab:

- A room or area that's large enough to set up test equipment and antenna stands next to each other, with up to 15 feet between the equipment and the stands and 10 feet on all sides around the test area. Ideally, the area is at least 30 feet by 45 feet and wide open. The bigger the better. With ceilings, higher is better, too.

- Minimal metal in walls and no cabinets or tables with significant metal content in the immediate area. (Metal studs are acceptable in plasterboard walls if you've got a room that is 30 to 50 feet wide.)

- No metal buildings or sheds.

- No metal ceilings, roof structures, floor structures, or wall structures within 30 feet of the perimeter of the test lab area.

- An area above and below the floor of the target lab area with the same attributes as the lab location itself.

- A clean RF environment.

Taking AEN into account

Think of setting up your lab in the same way you'd think about deploying a production RFID environment. Figure out how much *ambient electromagnetic noise* (AEN) is in the proposed area and what to do if there's a lot of AEN, which might cause interference. Chapter 7 takes you through the details of a Full Faraday Cycle Analysis (FFCA), which you need to perform before finalizing any lab location. You can either perform the analysis yourself with a spectrum analyzer or hire a firm to perform the analysis for you.

A laboratory environment needs to be much cleaner than a typical production environment. If an existing hand-held system or a security or location system is running over RF in a warehouse, it can usually work in concert with an RFID system. For a laboratory setting, however, you can't have any extraneous noise if you want accurate results. If you find any AEN where you're planning the lab, you need to either move the lab or eliminate the source of AEN. If you're lucky enough to have an unlimited budget, you can buy anechoic material or shielding and insulate the lab.

Considering power and network connectivity

A lab helps you create important data that will be useful in your production environment. To deal with that lab data and eventually to collect production data for analysis, make sure you have sufficient power and network connectivity. An important long-term consideration is the ability to share data between your production environment and your testing environment, so setting up a dedicated network for the lab is an ideal scenario.

I like to have a server cabinet in one corner of the lab that is close to the network connects (or "drops") that are built into the wall like electrical outlets. From the server cabinet, you can network the lab from the servers to the printers to the readers. Enclose the cabinet to protect the hardware from dust, heat, and impact. Most IT folks like to make the RFID network its own stand-alone network, separated from the regular corporate network, so there's no possible threat of intrusion.

Deciding between an office or a warehouse

When deciding whether to locate your lab in an office setting or in a warehouse, you need to consider the amount and size of the equipment you'll be using or testing:

- **Office setting:** If your lab is in an office setting, you probably can't construct a dock door portal and use a forklift, unless you've got a very forgiving CEO and landlord. An office setting is best for testing the products or SKUs in a static environment (one that doesn't use the movement and noise of the loop conveyor or Cyclotron), using frequency response characterization (which I explain in Chapter 9 on tag testing protocol). You can easily grid the floor and lay out a testing of readers and tags, but you won't have the benefit of validating your testing on a dock door or conveyor.

- **Warehouse:** If you want to test full pallets, stretch-wrap machines, and dock door configurations, consider putting your lab in a warehouse environment. The perfect scenario is to set up a lab in a nicely built-out portion of a warehouse. The build-out must be clean, open, and well-finished, including finished walls and either carpeted or vinyl floors to eliminate dust. The doors should be double-wide in case you want to move a full pallet into the finished area as well.

  For one Fortune 100 client, we built a clean testing environment in an office-like setup within the warehouse and networked it with testing stations on dock doors, shrink-wrap stations, and conveyors out on the warehouse floor. This is the ideal setup. The more individual components of a production facility (dock doors, forklifts, shrink-wrap stations) you can include in the testing environment, the better your information will be as you move from testing to pilot to production.

Building an anechoic chamber

An *anechoic chamber* is the bling-bling of the RFID world. It's a dedicated space where no RF waves can come in from the outside, and none of the waves your RFID equipment produces will bounce or echo to confuse your results. If you're lucky enough to have the budget to build one, you'll be the envy of your RFID posse. Building an anechoic chamber is easier than you might think, but it's definitely not cheap. A typical anechoic chamber includes 3, 5, or 10 meters of testing area. If you can dedicate a room roughly 20 feet long by 15 feet wide by at least 10 feet high, you're off to a good start. Here are some general guidelines for constructing it:

- Build the walls out of plywood rather than sheetrock.

- Line the outside of the plywood walls (and ceiling) with RF shielding to keep out any ambient electromagnetic noise (AEN) — like radios, cell phones, alarm systems, and

the like. The shielding material is usually metallic plates with electrical contact between them. Aluminum or copper panels make for the best shielding.

- Line the inside walls, ceiling, and floors with absorbing material. This material can be cone-shaped polyurethane integrated with carbon or tiles.

- Cover the floor with ferrite tiles to further absorb RF waves.

An anechoic chamber is a perfect environment in which to test tags, readers, and antennas and to find absolute data. It's a great luxury to have, but if you can't afford one, many universities and commercial companies will rent out time in their chambers. Either way, just knowing the term *anechoic chamber* makes you cooler than the other side of the pillow.

If you're looking for dedicated space for your lab in an office building or industrial park, be sure you know who your neighbors are and what they do *before* you sign a lease. For example, say that you find what seems like the perfect environment: a great 5,000-square-foot flex space with a big built-out clean front area and a large warehouse, dock door, and heavy power in the back. However, you later discover that the tenants next door include an airplane radar manufacturer and a machine shop — the next thing you know you're feuding like the Hatfields and McCoys. To protect your company from this scenario, make sure the facility goes through a 48-hour Full Faraday Cycle Analysis and passes with no AEN before committing to your lease.

Physics eye for the lab guy: Design the physical layout

Although you may not be the person in your house picking out the curtains (sorry, Honey — *window dressings*) to match the sofa, you need to be the person designing the layout of the laboratory environment, which is the next step after finding the right location. Just as the secret to good home decor is knowing the nuances of a color palette, the trick to designing a functionally efficient lab is knowing how to preserve open space.

Drafting your plan on paper

On a piece of graph paper, create a homemade floor plan for initial design and for follow-on experiments. Here are some general guidelines for creating this plan on paper:

- ✔ Lay out the floor of the test lab in a 1 foot = ¼ inch scale.

- ✔ Mark out power locations (and voltage) and *network drops* (connectivity points).

- ✔ Based on available power, plan out your main testing floor space in the middle of as much power as possible. If one end of the room has power and connectivity on all three sides, use that as your main testing area. Use the guidelines I mention earlier in this chapter to determine the size of the space.

Creating a grid pattern on the testing floor

After you earmark a large area of open space to serve as your main testing floor, you need to create the points of reference on the floor with a grid (like a giant version of your graph paper). This is where your 1-inch colored tape comes in or where you put your painting skills to use. The best way to create this grid is to follow these steps:

1. **Decide what will be the dead center of the testing space.**

2. **To create the grid pattern, first lay out (using tape or paint) a set of parallel lines off the center point.**

   I like to use 1 foot of separation between parallel lines. Making the lines parallel to a wall also helps because you can keep the lines straight by measuring an equal amount off the same wall at various points.

3. **After you have a good set of lines running parallel in one direction, lay down a set of perpendicular lines (exactly 90 degrees the other way) and create those lines across your original set.**

4. **Go back to your center point and strike two lines from corner to corner through the center of the grid so that you have lines at 45 degrees off center.**

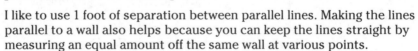

   The final result looks like the pattern shown in Figure 8-1.

Figure 8-1:
The floor
grid pattern
of your
primary lab
space helps
you record
accurate
test results.

Setting up areas for computer monitors

The next order of business is setting up workbenches and areas for various computer monitors. I say monitors because ideally you should have print-and-apply machines, readers, and test equipment hooked up to those monitors as well as to servers and desktop computers. Here are some general guidelines for setting up these areas:

- ✔ **Workbenches:** These can be something as simple as folding banquet tables or as sophisticated as built-in sturdy wood-framed workbenches with shelves and tool storage areas. Your imagination — and your budget — are the only limitations.

  When setting up your workbenches, make sure that they are located near ample power and situated along the edge of the lab so they don't take up any more floor space than necessary. If you can cluster your workbenches in two areas — one near the server rack and the other at the opposite end of the lab — two or more people can work at the same time without tripping all over each other.

- ✔ **Workstations:** Set up the different workstations by category, such as software/networking, hardware setup and configuration, and lastly a dedicated testing area. If the four corners of your room all have ample power and connectivity, put the server rack in one corner with the software/networking stations, put a reader setup and configuration area in the corner with the least connectivity, put a SKU testing station in another corner, and the last corner you can decide on as you discover your specific needs.

✔ **Bins, shelves, and other storage:** Next, fill the walls along the edges with bins, shelves, and other storage for tools and equipment. It seems like you can never have enough storage, and some things, such as antennas, require careful consideration when making storage areas. Figure 8-2 shows a simple solution for keeping various types of antennas well organized and protected. Cardboard shelves from an office supply store make perfect storage space for only a couple of dollars. If you want to protect certain storage areas from RF energy, you can build screens of any thin metallic material; however a screening solution also creates reflection, which may affect the rest of your testing environment, so be careful.

Figure 8-2:
Inexpensive and effective antenna storage from your local office supply store.

Set up the test equipment

After your lab is located, designed, and partially outfitted, it's time to put all that fancy electronic gear to work. The third step is setting up your lab gear. Before you set up the test equipment, consider whether you'll ever need to move the gear:

✔ **If you know something will be a permanent fixture in the lab,** set it up like a permanent fixture. Put it in a rack, tie down the cabling and wire neatly, and protect it from impact and grime.

✔ **If you know something will be used out on location (like spectrum analyzers and signal generators),** think about creating an area that is analogous to a docking station for a laptop. Pick a spot where the gear will be useful while in the lab but can quickly be removed and taken afield. See the sidebar, "Protecting your gear on the go," for details about protecting gear if you need to move it often.

In the lab, deploy testing stations that enable you to do both scientific testing and application testing:

✔ **Scientific testing** covers things like profiling RF interrogation zones out of various antennas, determining the variability of tag types, and tag production performance.

✔ **Application testing** covers things like understanding reader configurations for a dock door and building mock portals.

For scientific testing, you need enough electricity to power several readers and antennas on the very edge of your grid. For the application testing, you need access to power and network connectivity.

Reader testing station

A reader test station is where you can test antenna patterns on the grid, certify the power output and bandwidth range, and in general play with your readers while recording performance. Here are some tips for choosing a location for the reader test station:

Protecting your gear on the go

Any equipment you think might be needed out in the field, like signal generators and spectrum analyzers, should be protected while en route. The best bet is to buy a Pelican case and either custom cut the foam or have a spray-in-foam shop spray in foam so that the unit is well protected. Pelican cases can be checked baggage on a plane, shipped via most overnight carriers, and locked up when the gear is left in a warehouse. Consider it a couple hundred dollars of insurance for your costly lab gear. Buy the case when you buy the electronics, and you'll never worry about how you're going to get a case to Waterloo, Iowa, in a day when you're all the way out in Alamo, Texas.

✔ The ideal location is near the network gear, so that you can store and access important data easily.

✔ If one area of the lab has a concentration of power and network connectivity, that's where to put a server cabinet and create your reader test station.

✔ To get the best test results, put the readers at the far end of your floor grid if possible, as opposed to the middle. It doesn't matter which end of the room you choose, just as long as you can make use of the longest distance from one end of your testing area to the other by putting the reader at one end and a tagged object at the opposite end.

To set up the reader station, follow these steps:

1. **Set up the hubs and connect them to one or two dedicated servers.**

   These servers are the control PCs for the readers.

2. **With each reader, you get software to run the reader either as you would in production or for test purposes. Install the software on the reader server (a server that is dedicated to the readers and their control software).**

3. **Connect the hubs and servers so that if any of your readers need to be controlled, they can all communicate via the one dedicated reader server.**

   This allows you to connect multiple readers to the same server without having to plug and unplug each one.

SKU testing station

If you start with the grid, the servers, and the readers as anchor points in your lab, the next geographically located test station is the product or SKU testing station. This station consists of a reader or tag testing appliance, a single antenna, an antenna stand, an RF-friendly stand (made of dried wood or plastic) for the product to be tested, and power access. The floor needs to be measured, but being on the grid is not a requirement. Mark off the floor with 1-inch increments, but a full grid is not necessary. If the station is close enough to the server cabinet to run a couple of CAT-5 cables, that's perfect.

Conveyor

If you have a conveyor, you should put it as far away from the static testing area as possible. All that metal can have an unwelcome reflective effect on your testing. Using a Cyclotron (described in Chapter 10) is a better solution, but it still might cause enough noise to warrant separating it as far from the test grid as possible. The ideal is a room that's 75 feet long, which is big enough so that a loop conveyor at one end has a minimal effect on testing at the other end of the room.

Print-and-apply station

The last segregated area of your lab is the print-and-apply station. This station requires relatively little power and only a few points of network connectivity. You can run most print-and-apply solutions during other testing without interference because print-and-apply machines are such low-power output devices and are directionally focused.

As you begin to understand which areas need to be in close proximity to each other, you can move the areas around to suit your needs. In general, starting off with a lab design like the one shown in Figure 8-3 will get you off to a good start.

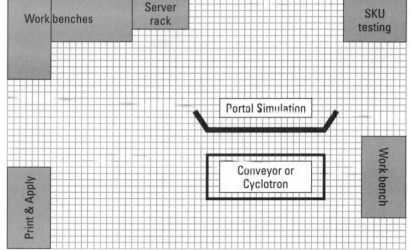

Figure 8-3: A basic RFID lab design and layout.

Build specific test equipment

The RFID industry is so new that you can't yet go down to your local Home Depot and pick up a couple of RFID antenna racks; you have to build them yourself. In the production environment, you have to think about protection, durability, and other considerations, but in the lab environment, you need to think about two things — portability and RF friendliness. Fortunately, the solution is cheap and simple. In the words of *The Graduate,* "Plastics."

You may wonder why PVC is on the list of must-have tools described earlier in the chapter. The reason is that PVC is the perfect material to make antenna racks, temporary portals, and conveyor stands. You can cut through it with a $5 handsaw, and you can put it together and pull it apart like LEGO bricks. If you add an assortment of connection joints, some plastic zip ties, and a bit of heavy-duty Velcro, you've got the makings of most of your lab gear.

The process of building PVC antenna holders and racks is pretty self-explanatory, so rather than give you step-by-step instructions, I'm just going to show you a picture of a design to hold a single antenna to give you the basic idea. Figure 8-4 shows the back of a PVC antenna mount on the testing grid.

You can make these PVC racks just about any shape or size that suits your needs. The important thing to remember is that you are trying to simulate the real-world environment. So if your conveyors are 12 inches off the floor, your test stands should be 12 inches tall. And if they really are that tall, you should let Mr. Claus know that you won't have any problems passing OSHA inspections, and the entire facility will be RFID-enabled by next Christmas because of your way-cool lab.

Figure 8-4:
A home-made, freestanding PVC antenna rack.

Using the PVC material for testing apparatus does have one drawback: It can't hold a significant amount of weight. To test cases of items or heavy objects for their RF suitability or for basic SKU testing, you need an RF-friendly material that also strong enough to hold up things like a 30-pound case of SPAM or a 75-pound uninterruptible power supply (UPS). For this, nothing works like good ol'-fashioned kiln-dried lumber. Lumber has minimal moisture content, but for purposes of SKU testing, not enough to affect your results. And whatever liquid the lumber has is a consistent factor. Figure 8-5 shows a sturdy, homemade, 2-x-2-foot SKU testing bench about countertop high.

Figure 8-5:
A SKU testing bench made of kiln-dried lumber.

Develop and implement standardized test procedures

When you're done setting up the lab, it's time to document your testing procedures. The purpose of documenting your testing procedures is to ensure that any new lab rat you hire can perform a test and compare the results with a test that someone else did six months or six years ago. In addition to being able to compare testing results, documentation enables new folks to get up to speed quickly on how things are done.

In your RFID lab, you can perform tests for an infinite number of purposes. However, to start with, perform the following four tests to quickly address some common issues:

- ✔ Antenna patterning test
- ✔ Reader performance test (for distance, speed, and accuracy)
- ✔ SKU test
- ✔ Tag characterization test

Antenna patterning and reader performance tests are performed in the area of the lab with the floor grid. SKU testing and tag characterization can take place in the area with the least amount of interference and likelihood of reflection or standing waves.

The four tests you perform in the lab directly correlate to the real-world environment. If you document your information well, you can take that data and use it to design and select your final system.

Antenna patterning test

This simplest test, perhaps the most insightful for those new to RFID technology, is to pattern an antenna and reader by following these steps:

1. **Set up a reader with factory default settings and attach a single transmit and receive antenna to the reader.**

   Mount the antenna to your PVC antenna stand or to a camera tripod at the edge of the grid. The top of the antenna should be roughly 4 feet off the ground, and if you dropped a plumb line straight down from the front face of the antenna, you should be right on one of the grid lines. See Chapter 10 for more on installing readers and antennas.

2. **Mount an RFID tag on a piece of cardboard attached to a tripod or PVC stand at the same height as the antenna that you set up in Step 1.**

3. **Start reading the tag at the grid line 1 foot away from the antenna. Begin reading the tag directly in front of the antenna and then move it in 6-inch increments to each side until the reader no longer registers a read.**

   The tag should stay one grid line away from the antenna the whole time, but you move it away from the center line.

4. **Record the location of the last successful read to get the outward limit of the interrogation zone at 1 foot away.**

5. **Now bring the tag back to directly in front of the antenna, but move it another foot away so that you're on the 2-foot grid line. Move the tag laterally along the grid line in 6-inch increments until you can't get any more reads, and then record the location of the last successful read.**

6. **Continue with each grid line until you can't get a read directly in front of the antenna or you run out of space.**

7. **Change the power settings on the reader and repeat Steps 3 through 6. Record the results and note the difference that varying the power level has on creating an interrogation zone pattern.**

When you're done, you have a graphical representation of an antenna pattern. The more places where you find that point between reads and no reads, the more accurate your pattern will be. You may also notice that something in your lab is distorting the perfect shape. If the shape of the antenna pattern is not symmetrical, look for potential causes, like a metal server rack or metal door frame. The only way to fix this is to remove the objects causing any potential distortion. Figure 8-6 shows a lab test result compared with the actual antenna field in a perfect environment.

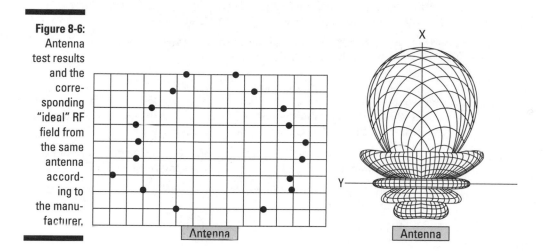

This test gives you a good understanding of what the interrogation zone or RF bubble will look like. With this knowledge, you will know what the pattern will be in the real world. With the data from your antenna testing, coupled with the RF path loss contour map discussed in Chapter 7, you can decide what antenna you should use and what the power setting on the reader should be. The antenna patterning test helps you design a much more accurate and effective system than you'd get by trial and error.

Reader performance test

Reader performance testing is the Super Bowl of lab work. It's exciting, it changes with every new piece of hardware that comes out, and after a while, you start to develop your own favorites. Chapter 10 gives you the complete rundown on how to test reader performance. The reader performance testing and the information about factors like connectivity and control discussed in Chapter 10 help you pick the optimal reader for your needs.

SKU testing

SKU or product testing for the optimal tag and the optimal placement has evolved into a science all its own. Here are the basic steps to SKU testing, which are covered in much greater detail in Chapter 9:

1. **Determine the five best locations to put a tag based on what you observe about the item.**

Is there a place on the case with nothing but air behind it, or some Styrofoam packing material rather than an actual product? Are there areas where the corrugated cardboard is double-thick, like where the box is sealed? Those are the first two things to look for — airspace behind where you'd affix the tag and extra thick cardboard. Refer to Chapter 9 for additional tips on where to place tags.

2. **After you determine your five test areas, figure out how many tags you want to test in those locations.**

   At a minimum, you should try four tags in each location.

3. **Set up a reader to separate the Industrial, Scientific, and Medical (ISM) band (902–928 MHz in most cases for U.S. applications) into individual channels and report successful reads on each channel.**

   All the readers can do this, but it takes some software programming to make the results meaningful, or a specific testing program.

4. **Attach a single antenna to the reader and put it 1 foot away from the product to be tested.**

5. **Record the number of successful reads divided by the number of tries. (This gives you the probability of the item being read.)**

If you don't have the software or the programming expertise, an easy, although much less scientific, way to test tags for their performance is to set up the reader to record successful reads and read the item for three minutes. Record the number of reads for each tag in each of the five locations you identified in Step 1.

Some commercial testing applications automate the tag testing process by breaking down the ISM band into 124 separate channels and then reporting back the probability of success for each channel. ODIN technologies' Trifecta software is the first software to automate this process and is available as a Web service, so you don't need to support another full software application on-site. This is a much more scientific approach than simply finding the number of reads over time.

6. **Repeat Step 5 for each tag location; then repeat Steps 4 and 5 at a 3 foot and 5 foot separation from tag to antenna.**

Figure 8-7 shows a screen shot from ODIN technologies' Trifecta testing software. The graph shows 50 of the 124 channels tested. The gray bars represent successful tag wake-ups, and the black bars represent successful reads. Black good, gray okay, nothing bad — got it? You can also see statistics of read success. This testing program is a simple automated way of testing the success of each tag and position. A program like this is particularly valuable if you have hundreds of different SKUs to test. Commercial testing labs charge between $1,000 and $3,000 per SKU depending on the quality of the testing and the usefulness of the data. It is easy to see how an in-house lab with some quality testing software can quickly pay for itself.

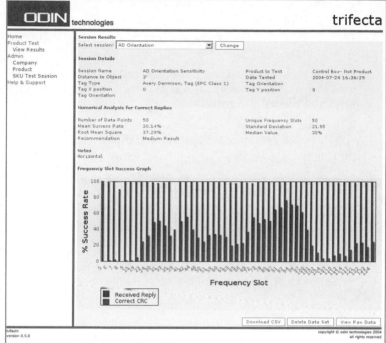

Figure 8-7:
A screen
shot from
Trifecta
showing
the perfor-
mance of
a specific
tag and
location.

The SKU or product testing is one step toward deciding what your end system will look like. Chapter 9 walks you through the specifics of tag selection and testing, but you have a few additional decisions to make:

- ✔ Do you want to get the best price from the tag suppliers by standardizing on one design and ordering in greater quantity?
- ✔ Do you need a tag that fits in with your print-and-apply machine?
- ✔ Do you want to use specialty tags for certain items?

The SKU testing gives you the knowledge you need to make these decisions.

Tag characterization test

Tag characterization is a relatively straightforward test that helps you determine the threshold of quality from various manufacturers. You need an attenuator for this test. Follow these steps to run the test:

1. **Hook up a reader to a single antenna, through an attenuator, and back into the reader.**

2. **Mount the antenna 2 feet away from a flat surface like your SKU testing stand.**

You need to mount the antenna above the stand and point it toward the ground.

3. **To characterize a certain type of tag, gather at least 100 of the same type of tags.**

4. **Test the tags one at a time under the antenna for readability. Slowly turn down or "attenuate" the signal until the tag can no longer be read and record the results.**

5. **When you have complete data for 100 tags, set the antenna up on the edge of your grid that's furthest from the direction you will be positioning the tag, so you can get as far away from the reader as possible.**

6. **Test several of the tags, one at a time, that were of the highest attenuation value and see how far they can be read and then repeat the process for each of the different tag attenuation values.**

The tag characterization test resolves one of the biggest issues you're likely to face if you use an automated print-and-apply solution: the lack of standardized quality from tag vendors. For example, a print-and-apply machine might be able to read a particular tag if it's a few inches away from the antenna, but might not be able to read it from a foot away or at a high speed because the chip is poorly mounted to the antenna.

The only way to tell the good tags from the bad tags is to characterize the quality of various tags and understand the lowest performance threshold that is acceptable given your mandates. In other words, if you know a tag attenuated to a certain point can't be read beyond a foot, you want to set up a verification station after your print-and-apply station (or work with the print-and-apply vendor) attenuated down to the lowest threshold acceptable. By using the tag characterization data to set up a verification station, you can stop bad tags from being shipped out to your clients and not getting read. If all your tags are read, you get paid quicker and with less hassle by folks like the Department of Defense (DoD), Wal-Mart, Target, and others.

Additional tests

The four tests that I describe in this chapter give you great insight into the performance and characteristics of RFID systems. You can add more tests as you put your creativity to work and as you graduate up the learning curve. Eventually, you will develop tests that are unique to your application or production environment. If you're tracking top-secret hard drives covertly, for example, you may want to test how antennas perform when they are behind sheetrock or how tags perform at different orientations. If you're producing roller skates, you may end up slicing up a standard tag and experimenting with ways of affixing the tag onto the boot. Creativity and repeatability are the keys to good experiments.

Chapter 9

Tag, You're It: Testing for Best Tag Design and Placement

W.C. Fields' constant search for the 25-cent cigar is second only to Linda Dillman's search for a 5-cent RFID tag. Dillman, as the CIO of Wal-Mart, has been hounded by suppliers implementing RFID who claim that the only positive ROI comes with a 5-cent tag. Although this might be true for companies that count their product margins in fractions of pennies, many current RFID implementations demonstrate acceptable ROI using today's tags, which range in price from 22 to 50 cents each. The trick is knowing the difference between tags and how they interact in your inventory environment so that — regardless of price — you can make the right decision for your organization.

Think of choosing the best tags for an RFID system as picking the right bat to play a game of baseball. (Without a bat, there's just not much point in playing, but you can't just go out and grab any ol' hunk of wood.) You need to evaluate the bat's attributes, try it out in batting practice, and see how it fits with your swing if you want to hit 'em over the fence. The same is true for tags — you have to test them in their working environment to get successful results from your RFID system.

The ideal RFID system automates any counting function such as shipping, receiving, picking products for shipment, asset management, security, and so on. It does this by capturing data *without line of sight*. Although RFID promises to achieve this automated, non-line-of-sight communication, in many cases this communication succeeds only through careful tuning of key system parameters, including tag selection and placement, reader selection and configuration, interrogation antenna selection, and a thorough understanding of the environment. This chapter deals with the *tag-selection-and-placement* parameter, specifically with testing tags on difficult-to-read items.

In this chapter, you find out how the right tags help you get better read ranges and rates and improve the efficiency of your RFID network. You also discover why different products make tags behave in unique ways and the implications of these unusual behaviors as you create your system architecture. A little bit of understanding of tags goes a long way, whether you do the work yourself or enlist a third party to do it for you.

Ready, Set, Test!

The biggest error many people make when setting off on an RFID program is thinking that a single solution will work well for every situation. Unfortunately, there is no silver bullet. RFID systems are always custom designed for both the environment and the properties of the item to be tagged. Getting the right tags for various products is a critical and ongoing process; you may find that the optimal tags and their placement change as you change inventory and as the technology evolves and prices change. The first step to finding the right solution is to test the items that you will be tagging (or to have someone test them for you). A good assortment of tags to test and a rigorous methodology get you the answers you need to build a good RFID network. The two primary ways of testing for proper tags and placement are

- **Application testing** (also known as **trial and error**) using a conveyor or dock door.

- **Scientific static testing** to evaluate the way a product is affected in an RF field.

A solid testing procedure, as in any data-gathering exercise, is required. Figure 9-1 shows the three critical steps for incorporating tag testing into your organization and the factors influencing those different steps:

- **Product assessment:** Determines the optimal areas to place a tag to be tested, so you don't need to cover every inch of every product with each different tag. You can base this assessment on a tester's experience using a method like the RF pyramid I show later in the chapter, or you can use an automated RF Visualizer to show a product's various RF-friendly areas. This step tells you where the best place to test is — essentially it helps you find the sweet spot.

- **Product testing:** After determining the five or six sweet spots on the product that look optimal for testing tags, you need to choose the tags you plan to test, document the test locations, conduct the tests, and then determine the best performing tag. For purposes of comparison, you should record the performance measures specified in a universal and scientific manner such as using the ODIN tag performance index (or TPI) for each tag, location, and orientation. See "The ODIN tag performance index" sidebar later in this chapter. In order to test tags effectively, you need an RF-friendly location without ambient electronic noise, testing hardware, and (to make life easy) testing software that produces efficient and easy-to-understand results across the entire frequency band.

✔ **Employee training:** Incorporates tag testing into your normal inventory processes by making sure that you have a system in place and documented procedures that can be taught to end-users simply and concisely. Manufacturing employees or packaging workers are the likely recipients of your product-testing training, so keep their needs in mind when designing your program. And create a user's manual for your testing process that appropriately addresses this intended audience. Chapter 13 discusses user training in more detail.

Tag testing must isolate the performance of the tags for a given orientation (vertical, horizontal, random) and location on the case (top, back, side). Tag testing should not be a test of reader performance, antennas, or any other variable. *The tag is the only thing that should vary during the testing.* Keep this in mind when designing your testing methodology because the software on readers can vary results greatly, and some automated methods don't test beyond one or two channels in the ISM UHF band. (There are 124 channels, and you need to know the performance across the *whole* band.)

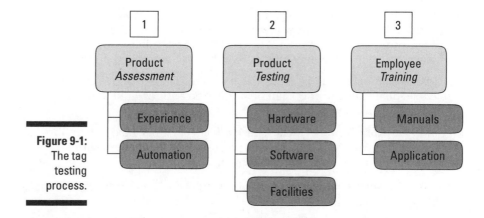

Figure 9-1: The tag testing process.

Recently, my company demonstrated the use of Trifecta for a client. Trifecta is an ODIN technologies software tool designed to simply and accurately provide a tag performance index so tags can be compared universally across the frequency band. I tested one of the client's difficult-to-read SKUs, which was attached to an item with a high water content. The client wanted to know why the RFID system was performing poorly. The problem turned out to be that the tag was not positioned in the optimal location. The client assumed that because the interrogation antenna had line of sight with the tag, he would instantly achieve perfect performance. I explained that with difficult-to-read objects, tag testing has to be done in order to be successful with RFID. The software helped me locate the optimal placement and orientation, illustrating that moving a tag even a few centimeters in one direction or another can make a huge difference in performance. Furthermore, I used the ODIN TPI to compare results from various locations and show that careful testing (leading to better tag placement and orientation) could improve a 30 percent successful-read result to a workable 82 percent result.

Looking at the Material Composition of the Items You're Tagging

Before you begin the actual testing, you need to consider some of the parameters that impact the behavior of the tags and their response to the frequency you're using. Here are the four principal effects that material can have on an RF signal and, therefore, on a tag:

- **Absorption:** Some materials absorb the energy of the direct wave propagating out of the reader's antenna. This situation is also known as *loss*, a term accurately describing that, with absorption, there is less power available to get a signal back from the tag.

- **Reflection/refraction:** Ideally, the tag receives a direct wave from the interrogating antenna, but sometimes, material around the tag can reflect or refract that direct wave. Then the tag receives the reflected or refracted wave, which may look different than the original direct wave.

- **Dielectric effects:** When a dielectric material is close to the tag, the electric-field concentration can be multiplied and result in a detuning effect on the tag's antenna.

- **Complex propagation effects:** These effects exist because of two phenomena that work together in your RFID system to interfere with its successful communication. The phenomena are

  - *Standing waves,* or waves different from the direct wave you're trying to get to the tag.

  - *Multi-pathing,* which is caused by the standing waves and can cancel out the direct wave altogether.

> The direct wave going to the tag from the reader antenna is a uniform wave. If another wave of the same size (a standing wave) is coming in the opposite direction exactly half a wave ahead or behind the direct wave, it will cancel out the direct wave.

Many materials being tagged today are packaged in corrugated cardboard boxes. Cardboard is permeable to RF waves and, therefore, has little effect on the behavior of the RF waves. But the material makeup of what's inside the box has a direct impact on how the RF waves behave.

Examining RF transparent, reflecting, and absorbing materials

Objects can exhibit a wide range of behavior characteristics in relation to RF that is dependent on their material composition. An object can be RF transparent, RF reflecting, or RF absorbing. Most objects exhibit some combination of the three. The following list examines the behavior of two common material types:

- **Metallic:** Metallic items are the most likely to be *RF reflecting*. The tag does not absorb enough power from the reader because the metallic item either shields or detunes the tag's antenna from its resonance frequency so that the antenna cannot absorb enough energy to power up. Metal is a difficult item to tag, but tags placed directly on metal can work well if they are specifically tuned. It is possible to use a metallic item as a *backplane,* or part of the antenna itself.

- **Liquid:** Liquid materials — like water, shampoo, saline solution, and the like — are *RF absorbing.* They absorb RF waves and eat up all the potential energy a tag needs. They reduce the strength of the original signal by absorbing or dissipating the power, again causing the tag to have insufficient energy to power up and backscatter information to the reader. However, not all liquids are created equal: Water reacts much differently than something like oil, for instance.

Table 9-1 shows the behaviors of different types of materials and their effects on an RFID communication system.

Table 9-1	Material Effects on RFID Communications
Material Composition	*Its Effects on RF Signals*
Corrugated cardboard	Absorption from moisture
Conductive liquids	Absorption

(continued)

Table 9-1	Material Effects on RFID Communications
Material Composition	*Its Effects on RF Signals*
Glass	Attenuation (weakening)
Groups of cans	Multiple propagation effects; reflection
Human body/animals	Absorption; detuning; reflection
Metals	Reflection
Plastics	Detuning (dielectric effect)

Different product attributes cause RF waves to behave differently in an RFID interrogation zone. The material properties of an item — the amount or content of metal, liquid, cardboard, airspace, and so on that the item possesses — all affect the way that the RF waves react around that item. The best way to think of it is in terms of sound waves. In an all-metal room, for example, you hear echoes and strange noises because the sound waves bounce off the metal. In a sound studio with cushy material lining the walls, sound is deadened as soon as it hits the walls. RF waves behave in a similar manner.

A combination of metallic and liquid materials can cause the items being tagged to absorb, shield, and reflect RF waves in a variety of different ways. For example, think of a case of salsa, a heavily liquid product stored in glass bottles with metal lids. All sorts of funky things happen in the RF world when you're tagging salsa. That is, reflected waves bounce off the metal, waves absorb into the liquid, and a general calamity of propagation effects make tagging a real challenge.

What you need to be concerned with is how those RF waves behave around the tag and how the materials close to the tag affect the waves' behavior. Figure 9-2 illustrates how two different cases of product behave in an RF field. The product on the left has a consistent and uniform behavior in the lines representing the RFID reflection and is representative of a case of paper towels. The product on the right is representative of a case with liquid absorbing the waves, glass attenuating them, and metal caps reflecting them. A product like the one shown on the right is much more challenging to tag because there are fewer areas where the tag can effectively couple with the signal. The dips in the lines around the case are called *nulls,* where there is little or no RF energy. A tag placed near one of those nulls will not read effectively.

Using the RF friendliness pyramid to understand the optimal spot for testing

I was fortunate enough to be part of the RFID Expert Group (REG) for AIM Global (The Association for Automatic Identification and Mobility), and we came up with the RFID friendliness pyramid as a way of representing properties of cases that contain various objects. Figure 9-3 shows the pyramid.

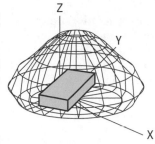
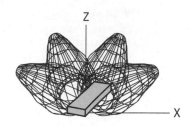

Figure 9-2: Two cases of different material in an RF interrogation zone.

The RFID pyramid is useful for determining how things should be tagged and, to a certain extent, where. The higher up the pyramid you go, the more stable the RFID performance is. For example, a cardboard box full of bubble wrap, which would be classified at the top of the pyramid, is easy to tag and read. The peak of the pyramid implies that the material is RF-transparent and is similar to reading a tag in free air.

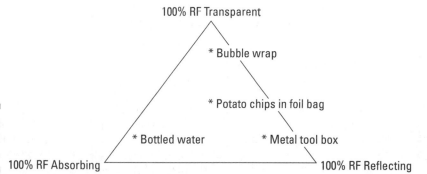

Figure 9-3: The RFID pyramid.

You can also use the pyramid to help you determine the optimal location to begin testing. In a typical case, you should be able to choose the top four or five locations to test based on basic knowledge of the laws of electromagnetism discussed in Chapter 5. If the object you want to tag is made up of many items with various RF properties, you can rate the various parts of the object based on their RF friendliness by using the pyramid.

To determine the best spot for testing, follow these steps:

1. **Determine whether the outside packaging material is RF transparent.**

   If the packaging material is a corrugated cardboard box, it is RF transparent. If the packaging contains metal, it's not RF transparent. You need to use packaging that is mostly RF transparent to get successful reads, or you need to use tags specially made for metal.

2. **If the packaging material is RF transparent, open the container and visually inspect the contents.**

   What you're likely to find is a mixture of reflective and absorbing materials. For instance, if you open a case of deodorant, the label on the deodorant may be metallic, surrounding the bulk of the product, which is a liquid-based aluminum oxide material. The caps and bottoms are all plastic and free space, and there may be some packaging material at the top to protect them.

3. **Determine where the contents fit on the RF pyramid.**

   For the deodorant example, the metallic labels would be in the lower-right corner of the pyramid because they are likely to be very reflective. The combination of liquid and aluminum would be in the middle near the bottom because it is both absorbing and reflecting.

4. **Rank eight or ten different areas of the object on their positions in the RFID pyramid based on how close to the peak they are.**

5. **Start your tag testing by using the highest ranked (most RF-friendly) areas.**

   For example, on the case of deodorant, start the testing with the plastic caps or bottoms, near the top or bottom of the case. The top would also likely have some air space and provide a good spot for placing a tag.

Anyplace there is extra cardboard material (like where the box top gets folded over and glued on a typical case) is also likely to provide extra insulation from any RF unfriendly material on the inside.

There is no silver bullet in RFID. Vendors who promise "RFID-in-a-box" or an instant slap-and-ship solution are selling you short. Many people have made the mistake of buying one of these solutions that may have reader x, tag y, and a certain middleware component. When they install the system and slap a tag somewhere on an item, particularly if the item is RF unfriendly, the system doesn't work. When this happens, people tend to blame the technology and say that RF is immature. The truth is that the users do not have a fundamental grasp of the physics of RFID and have not tailored a solution for their needs. Only by first assessing the environment and then testing the product to be tagged can you choose the right readers and tags.

Choosing a Tag to Test

Tags are not all created equal. In my testing of all the major tag manufacturers, I've found a wide (and wild) variance in production quality from tag to tag and from batch to batch. When I do any kind of tag testing, I make sure that I'm using a *normal* tag — that is, a tag that's squarely in the middle of the known performance of the tag type, not one that performs poorly or one that is extra-sensitive. To find out what average is, I use the link margin test developed at ODIN technologies labs to qualify tags as normal. Communications *link margin* refers to the amount of power that can be removed from the RF signal before communication is no longer interpretable.

Some tags couple with RF fields better than others. You can quantify a tag's RF coupling capacity and power efficiency by gradually reducing the strength of a readers' signal and recording the amount of power that must be removed to disable the communications channel. This link margin test quantifies how tags of the same type compare at extracting, consuming, and reflecting RF power. Tags with high link margin thresholds require less power to operate and therefore display superior read range and reliability.

To perform link margin testing, you need a device called a radial attenuator or software that can reduce the power automatically coming out of the tag. The *radial attenuator* is used to manually reduce the power in the receive path of the reader. You connect it to the transmit (Tx) coax cable between the circularly polarized interrogation antenna and the reader. This attenuator reduces the amplitude of an RF signal without affecting the waveform's phase, frequency, or shape in a significant way. The result is a direct relationship between attenuation and link margin: the higher the attenuation needed, the larger the link margin and the stronger the tag.

To get a statistically significant result, you need to test 100 tags of each type. It's easiest if you set up an antenna a few feet above a table where you put each tag to be tested — kind of like a big light hanging above an operating table. After setting up your test system (the radial attenuator between the interrogation antenna and the reader), follow these steps to perform the link margin test for each tag:

1. **Activate the reader and set the attenuator to its lowest output level.**

2. **Place a single tag in the field generated by the reader in autoread mode.**

3. **Manually adjust the attenuator to increase its power drain on the signal returning from the tag.**

   As the attenuation increases, the read rate gradually diminishes to a specific threshold, which you need to record for each tag.

4. **When the tag no longer can be read, record the corresponding value on the attenuator for the tag you're testing.**

   Use a permanent felt-tip pen to number the tag and write down the tag number and the attenuation value on your test record.

5. **Repeat steps 1 through 4 for all 100 tags.**

6. **Calculate the average attenuation value and standard deviation for the 100 tags tested.**

   The easiest way to track and analyze the data is to set up an Excel spreadsheet with the numbers 1 through 100 in their own cells descending in a column. To the right of each number, put their corresponding attenuation value. This enables you to highlight all the attenuation values and use both the average function, =average(data cells), and the standard deviation, =stdev(data cells). The average attenuation value becomes the basis for selecting the tags to use in your application testing.

7. **Go back to your stash of tested tags and pull out a bunch of tags that have a tested attenuation value that is the same as the calculated average. Also, record the standard deviation for the different tags you are testing.**

   The higher the standard deviation, the lower quality the tags and the more variability there will be in their performance.

   These chosen few are your application test tags. Now you're guaranteed that you don't have tags that are way better or way worse than the average tag you can expect.

Testing Tags in an Applications Test Facility

Many end users and integrators have developed *applications test facilities*. Typically, these facilities include

- ✔ **A circular conveyor equipped with multiple read stations:** Cases of the product being tested are tagged and carried around a conveyor at various speeds. During their journey, the demonstration software provided by the reader vendor records the total number of successful reads executed by each reader. The test process repeats for each tag and each position of interest to the client. Using this testing method is acceptable only if you know (and duplicate) the exact setup where the product will actually be read, which, based on the simple rules of physics, is almost impossible to do. The configuration, reader type, antenna choice, conveyor type and construction, and software all affect read rates.

- ✔ **The dock door simulation:** Typically, four interrogation antennas are mounted on a portal measuring about 10 feet x 10 feet. A pallet of tagged product moves through the field, and the test equipment setup records the total number of successful reads. More advanced labs have software that depicts not only the total number of cases read but also where on the pallet the cases were read, making pallet optimization more efficient.

The ODIN tag performance index (TPI)

In the past five years, the Auto-ID lab at the Massachusetts Institute of Technology (MIT) has led the way in testing and design of RFID systems, and ODIN technologies has extended that research and testing into the commercial world with the help of the current director of the Auto-ID Labs, Dr. Daniel Engels. Both organizations and many end users have a mutual goal of creating a common method for testing tag performance and comparing results. ODIN technologies is leading the effort to create a specification that the industry can talk about with common vernacular. This specification is called the *tag performance index,* or *TPI.*

The TPI is a method of gauging tag performance via three critical performance measures across all channels in a particular frequency band (124 channels can be isolated in the 902–928 MHz band, for instance):

✔ Percent of successful reads

✔ Dispersion of reads across the ISM band

✔ Ratio of nulls, or poor performers, to successful reads

After these measurements are taken, they are entered into a simple formula that yields a TPI, which can range from 0 to 100. The TPI is a proxy for how well a tag will perform in a real world situation. It is the only effective way of knowing the effects of de-tuning by material type on tag performance.

The TPI is useful for several reasons, but most importantly, it identifies the threshold of performance to meet particular application requirements. For instance, a retailer can use the TPI to create a mandate for its suppliers. Rather than issuing a non-scientific or non-measurable standard — such as, the pallet tag needs to be read going through a ten foot dock door or on a 600 fpm conveyor — the retailer can require that all cases have a tag and tag placement combination TPI measurement that is equal to or greater than 70.

The TPI can also allow tag manufacturers to create a common way of comparing tag performance across varying frequencies and chip designs, and on differing products. The use of this measurement will further promote packaging designs with a known TPI specification.

The other methods used to test tags — application testing and power attenuation — fall short of creating a system that accounts for performance across a wide band of frequency, works in any country at any frequency, ignores reader firmware and power, and is affordable and accessible to all end-users. The TPI creates a spec that economically allows users to test multiple combinations and permutations and is grounded in a very scientific methodology.

Because no single standard reader setup, conveyor, or dock door configuration covers every use, each application testing facility will be a little different. A standardized testing process is difficult when there are so many different random variables. These factors make application testing a poor way of testing, and you should use it only for validation of other test results because a battery of tests is more accurate than one specific test. Here's another drawback: If you're testing with your own setup, you can't call your buddy at a different company and compare your results.

The sections that follow give you the lowdown on how to carry out tag testing in a more scientific manner. In order to get the optimal performance when using an application test, you should follow a structured process, even if you

don't have a full-scale lab. Using ISO methods 18046 and 18047 for confor-
mance measurements is helpful. The ISO method is important to make sure
that you can repeat your testing results and that you have a documented procedure. The problem with application testing, however, occurs when you
can't really isolate one variable to be tested because you can never be sure
what might be causing the difference in performance between one test and
another.

Setting up the testing environment

If you have a large open area (either indoors or outside) with little or no
ambient electromagnetic noise (or AEN, as determined by the testing
methodology in Chapter 7), you can set up your tag testing environment in
that area. Here's the equipment you need to set up the environment:

- **A fixed-mount RFID reader and single circularly polarized antenna**
  attached to a tripod or antenna holder like the ones shown in Chapter 7.
  The reader should be connected to

- **A tag testing software capable of characterizing the frequency
  response at each channel in the band you are testing (most likely
  902–928 MHz)** loaded onto a controlling PC with software.

- **A wooden or plastic table** suitable for holding the items to be tested.
  The table should be the same height as the interrogating antenna.

- **A tape measure** for marking out several distances.

Follow these steps to set up your test system:

1. **Position the antenna and reader at least 25 feet from any obstructions
   that may interfere with the interrogation zone.**

2. **Mount the antenna approximately 3 feet off the ground in a fixed
   location.**

3. **Mark off 1-foot measurements heading directly out from the antenna.**

4. **Position the table with the tagged product directly in front of the
   antenna, 1 foot away.**

   You then move the tagged product sitting on the table out to 3 and 5 feet
   to compare tag performance.

Carrying out the test

You measure the effect of different placements and tag types as they are
affected by the internal contents of the product, or SKU, you are testing. You
can do this in three ways:

✔ Use the tag testing software to gather a frequency response characterization (FRC), described in more detail later in the chapter, at the three different distances — 1, 3, and 5 feet. Then calculate a tag performance index (TPI) as discussed in the sidebar.

✔ Move the tagged object farther and farther away until it cannot be read and then compare different distances with each other.

✔ Take reader threshold power measurements (explained below). This helps overcome the issue of reader software, which can vary from reader to reader.

Using the reader power threshold is a good way to determine the quality of the tag and tag placement, but results can vary from reader to reader and location to location. It is a good method if you can do all your testing at the same time and only need to know relative performance, but keep in mind different readers will affect the results in different ways. To do this, start with the maximum power and then monitor the percent of good reads as the power is slowly and incrementally reduced:

✔ Record the power threshold value in dB at full power. Record the same number of reads per minute — or if you are testing on a conveyor, reads per pass — that you get for each successively lower dB increment. Record the final dB value where you were able to get a successful read.

✔ In a given test, the best result is the lowest power level you can obtain and still achieve accurate read results.

✔ Map out the results of your testing in a matrix (or table), which shows the location of the tag and the results from the testing. I use the upper-left corner as the 0 point on an *x-y* axis. This allows simple comparison of data.

Frequency Response Characterization: Testing Tags with Physics

A more scientific approach than a specific application test (described in the preceding section) revolves around

✔ Understanding how an RFID reader interrogates (communicates with) a tag

✔ Being able to record performance of specific tags by using those communication principles that drive reader performance

✔ Eliminating all the variables that affect read results except the tag type

RFID readers *hop* from frequency to frequency within the UHF ISM band (902–928 MHz in the United States) using a pseudo-random sequence. (If you've ever watched a physicist dance, that defines *pseudo-random*.) By hopping randomly between frequencies, it's unlikely that two readers will try to communicate on the same frequency simultaneously, thereby avoiding interference. It's kind of like when your spouse keeps your mother-in-law out of the garage — no interaction means no interference.

However, due to the material properties of both the tags and the items (see the section "Looking at the Material Composition of the Items You're Tagging," earlier in the chapter), each product/tag combination may have a frequency preference. In fact, some product/tag combinations simply don't work in certain frequency channels. So identifying a combination that works well across the entire RFID spectrum of choice is important, which is precisely what *frequency response characterization* does. Rather than relying on large pieces of hardware to move products through an interrogation field, frequency response characterization uses intelligent software to control the single parameter that matters most in RFID: the *radio frequency*. (Hey, you know it's important because this variable constitutes half the acronym.)

It is possible to build such a software tool yourself if you understand the interface of a signal generator, spectrum analyzer, or reader, and if you have experience with RFID readers and coding principles. However, it is much simpler to invest in a commercially available tool like ODIN technologies' Trifecta or to have a certified test lab figure out the characterization of each of your products. If you decide to build a tool yourself, the critical component is the capability to isolate the reader's performance and test it to statistical significance over each channel in your chosen band. You can then compare the results from each test at various distances to determine the best tag and location. Table 9-2 shows the results of an internal test across all the channels, comparing three different tags.

Table 9-2	Comparing Tags at Three Distances			
		% Reads @ Distance		
Tag Type	Placement	1'	3'	5'
I Tag	1	98.36%	68.42%	24.64%
	2	97.36%	17.72%	1.78%
	3	98.54%	1.12%	0.00%
Strip	4	98.10%	40.18%	4.54%
	5	97.92%	3.76%	0.00%
	6	97.92%	1.10%	0.00%

Tag Type	Placement	% Reads @ Distance		
		1'	3'	5'
Squiggle	7	99.46%	71.40%	27.58%
	8	95.72%	71.74%	22.15%
	9	95.72%	25.48%	0.00%
	10	99.36%	5.38%	0.00%
	11	95.64%	77.82%	14.22%

Trifecta, from ODIN technologies, is an example of a commercially available tool that can perform this type of test. It uses frequency response characterization to discover the perfect tag selection, position, and orientation. The testing protocol consists of placing a stationary product in front of the interrogation antenna connected to a Trifecta server. The user simply clicks a button in the software to launch a test, which breaks down the RFID spectrum into 124 different channels, picks 50 channels at random, and issues 100 read commands on each channel. A few seconds later, the test is complete, and Trifecta produces the statistical information shown in Figure 9-4, reporting the number of successful read cycles executed on each channel (lighter gray bars) and the number of successful tag wake-ups (darker gray bars), which quantifies tag performance across the entire RFID band.

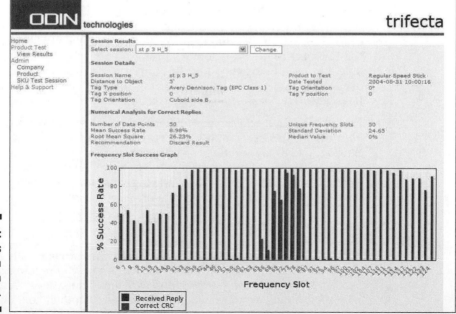

Figure 9-4:
The results of a Trifecta test on a product.

Encoding and Applying Tags

After you determine which tag position is best for each of your SKUs, it's time to figure out how to encode and apply your tag to a case or pallet in your production environment. This is an important question that impacts existing business processes, physical layout, and information technology systems. The next few sections detail some common approaches to encoding and applying tags.

Tag and ship

RFID technology is foreign to most companies currently working to comply with an EPC mandate. For this reason, many companies have opted to forego any potential return on investment in pursuit of EPC compliance alone. As such, *Tag and ship* represents a minimalist approach to RFID. It typically involves little to no software integration with existing applications and trades off a large capital investment (driven by hardware *and* software) for high labor costs into the future.

Tag and ship often takes place in the staging area of a warehouse. Several options exist for how pallets are unloaded, how tags are encoded and applied to each case, and how pallets are rebuilt.

Your options for encoding tags are as follows:

- ✔ **Preapplication:** The most common method is to use an RFID-enabled bar code printer to encode the data on the tag before it's applied to a case or pallet. These printers automatically test each RFID tag and write the EPC data to the tag before it's printed. This is by far the most reliable approach to tag encoding. It is important to select a printer that uses a communications protocol that is compatible with the tag you have selected (Class 0, 0+, or 1, for example).

- ✔ **Post-application:** You can use an RFID reader to encode data on the tag after the tag has been applied to a case or pallet. This is risky, however. Writing to RFID tags requires significantly more power and time than reading them, and the materials in the field can have an adverse effect on the reader's ability to write to the tag. The time constraints can also have an adverse impact on current conveyor speed.

You can apply tags in the following ways:

- ✔ **Manual application:** Many companies have opted to manually apply EPC tags to the products they ship to customers who require them, especially when the product is shipped in small quantities. ***Beware:*** Your SKU testing may reveal that some tag and product combinations require high-precision application. I've worked with SKUs that show dramatically poorer performance based on a ½-inch difference in position, and warehouse workers are not commonly known for this degree of precision. Having tags applied manually is also sometimes referred to as *slap and ship* for the imprecision of the method.

✔ **Automatic application:** For the tag/SKU combination that requires high precision or for high-volume throughput, automatic applicators are the right solution. They consistently place labels within a 1-mm target and integrate tag-encoding functionality as well.

These are your choices for rebuilding pallets:

✔ **The manual approach:** Workers manually take cases off the pallet and place them on the floor. Then they manually apply the tags and rebuild the pallet.

✔ **Reversible conveyor:** Some companies manually depalletize products onto a reversible conveyor (usually "S" or "U" shaped). After RFID tags are encoded and applied (manually or automatically), the conveyor is placed in reverse, and the pallet is rebuilt in the reverse order it was loaded onto the conveyor. This reverse process makes sure that the pallet is built the same way it was originally packed. To minimize the required floor space, the conveyor can take on a zigzag shape like the one shown in Figure 9-5, which I deployed at a warehouse in Texas.

RIFD racks housing readers and antennas

Figure 9-5:
An "S"
shaped
conveyor
with RFID
tagging and
verification
stations.

Inline production application

Tag and ship (described in the preceding section) is labor-intensive, so unless you ship products to your own warehouse, Tag and ship is merely a cost of doing business with no tangible return on investment (ROI — for the bean counters). The best way to begin testing and potentially realizing an RFID-based ROI is for the tag application process to migrate upstream into the production line. This technique offers the following benefits:

- ✔ Eliminates the need to break down and rebuild pallets in the warehouse staging area

- ✔ Supports the possibility of automatically tracking work in process

- ✔ Helps create better controls for companies fulfilling orders from inventory and then manufacturing to those order volumes

- ✔ May bring greater speed and accuracy to quality assurance and shipping processes

Despite these rosy attributes, you have to overcome several challenges as well:

- ✔ You must integrate an automated applicator into an existing production line, which requires some space on the manufacturing line.

- ✔ It's likely that you'll have to tag every product coming off the line, unless each line is reconfigured on a job basis (by running 1,000 cases at a time for a Wal-Mart order, for example).

For this approach to make the most sense, it's important to negotiate with the customer who has issued the EPC mandate to tag full pallet loads of a few SKUs only, rather than mixed pallets composed of many SKUs. Many companies have been successful in this approach. Although after just a few years, all products are likely to have EPC tags affixed to them and this distinction will become irrelevant.

Inline production software integration can occur on many different levels. The primary system's goal in a pilot situation, which inline application makes possible, is to gather valuable statistical data about the potential impact on standard warehouse operations. (For example, RFID might reduce picking time by 30 percent, streamline put away by 15 percent, and reduce order lead time by 12 hours.) Chapter 3 talks about the various business processes that may benefit from an RFID system. Conducting a pilot RFID project provides a firm foundation on which to base the business case for a more extensive rollout and integration of RFID infrastructure with existing legacy software systems.

In Chapter 15, I walk you through a nine-step approach to business case analysis. This analysis is much more accurate when supplied with real-world data that you learn from first piloting an RFID program in-house. Rather than installing RFID readers at all 30 dock doors in a warehouse, for example, you can take a more measured approach. By installing one reader at a receiving door and one on an outbound door, you can probe the potential of RFID and EPC compliance without making a huge capital investment. This approach also enables you to make the necessary adjustments to existing business processes and give your executives the opportunity to become comfortable with the large price tag of a full-scale deployment (usually several million dollars by the time data integration is considered).

The Secrets of Read Success

Techies are renowned for feverishly scribbling down notes on napkins when they have great new ideas, only to find them unreadable a day later when pulled out of their jacket pocket. Don't let this happen in your RFID deployment! Just because you've discovered the best tag and position for your product does not mean that your reader system is properly configured to read it. The following sections explore some common problems and tried-and-true solutions to ensure maximum read success.

Avoiding cross talk

Cross talk occurs when tags are detected by two or more readers responsible for monitoring different physical areas. This causes confusion when the data must be correlated to a specific dock door. In general, cross talk is caused by excessive signal strength, poor tuning of interrogation antennas, and poor reader configuration. Reading the same tags from multiple dock doors is typically unacceptable and must be designed out of the system.

Readers have varying degrees of *tunability,* which refers to a designer's ability to program a reader's performance characteristics. Fortunately, Generation 2.0 of the EPC protocol incorporates a special technology for a dense reader environment and eliminates many issues around tenability. Unfortunately, Gen 2.0 readers are not likely to be available until the end of 2005. In the meantime, here are some ways you may be able to avoid cross talk:

✔ **Adjust the power output.** Readers often provide the capability to adjust power output, thereby tuning the shape of the field emitted by interrogation antennas and preventing interaction with adjacent systems. Unfortunately, some materials require strong signal strength to penetrate and read the inner cases of a pallet. In these instances, you must use some other method to shape the field and avoid reader cross talk.

✔ **Utilize reader triggering.** This approach is more expensive and complex. Generally, a light beam is used to detect motion through a dock door. When motion is detected, the reader is activated, and the tag data is captured. This approach eliminates unnecessary radiation in the environment by powering down the field emitted by the reader when it's no longer needed. This technique typically adds $100 to $300 in hardware costs and often involves integrating a programmable logic controller (PLC).

✔ **Use software to coordinate readers.** In Chapter 11, I talk about the attributes of various middleware providers. Some of these companies are working on products to coordinate multiple readers, phasing their transmit and receive timing so that readers in close proximity do not interfere with each other. As of this writing, these products are not in a production state, but they do hold promise of making dense reader deployments easier.

Ensuring high-speed reads

Certain mandates require that tagged cases be readable on a conveyor at speeds up to 600 feet per minute (fpm). This rapid rate can be particularly challenging for difficult-to-read objects. Here are some ways that you may be able to solve this problem:

✔ **Tune the reader acquisition mode.** Readers provide varying degrees of control for how a read command can be structured. In a high-speed application, it is important to ensure that the *wakeup* and *read* commands are both issued rapidly, rather than just a few wakeups and many reads.

✔ **Change how the antennas are pointed.** A common mistake is to have antennas facing one another across a conveyor. Instead, they should be pointing 25–45 degrees relative to the conveyor and in the direction of conveyor motion. In this configuration, each case is in the field for a longer time, providing more opportunity to execute a successful read command.

Executing full pallet reads

Although no mandate currently requires 100 percent case reads on a pallet, some internal business cases require it. It really is the next logical step for

RFID, but some products seem nearly impossible to read on a pallet. Here are some suggestions on how to execute full pallet reads:

✔ **Place tags on the outer cases of a pallet.** When tagging cases, place the tags on the outer cases (relative to where they are stacked on a pallet) whenever possible. Also, try to leave an air gap between cases of product to allow deeper penetration of RF into the pallet.

✔ **Adjust antenna sequencing to focus RF power on challenging areas.** Most readers support four antennas to transmit their commands and receive data back from RFID tags. These antennas are never active simultaneously; instead, they are *multiplexed,* or switched in a programmable sequence. Readers provide varying degrees of flexibility for antenna sequencing. Some provide a customizable order, repetition, and power level for each antenna, making it possible to concentrate RF power on a given corner of a pallet where penetration is consistently challenging.

✔ **Tune the reader acquisition mode.** In this situation, many tags are in the field simultaneously and data collision becomes a potentially serious problem. In this case, it's important to issue many read commands for every wakeup to ensure the data is correctly received.

Chapter 10

Hooked on Phonics: Reader Testing, Selection, and Installation

In This Chapter

▶ Deciding whether to use hand-held, mobile, or fixed-location readers

▶ Identifying the costs involved

▶ Testing readers for performance

▶ Determining your connectivity needs

▶ Evaluating readers for built-in flexibility

▶ Installing readers and antennas

*I*magine that you had 12 months to find a partner and get married. Then imagine that you could go to a spouse store and view a whole lineup of potential mates, with all their good qualities listed beneath them. How easy would it be to pick one for life? If you could walk in and say that you want a blonde who speaks Italian and can cook Thai food, how do you think you'd end up? How would you even know what attributes to start looking for in the first place? And wouldn't it bother you that none of their negative attributes are listed? I bet you'd be bothered if your new mate turned into a giant green ogre at every sunset. You'd at least want to know ahead of time that you might need a bigger bed.

Although it may not be a permanent decision made "'til death do you part," choosing the right RFID reader is one of the most critical decisions for the success of your RFID network. It is something that is largely underrated, particularly by those trying to sell a middleware solution. All readers, like all spouses, are not created equal. If you need to comply with RFID by a certain deadline, your first step is to find out what you like in a reader. This chapter explains some of the practical attributes of RFID readers and what the implications of those attributes are for your system decision. I also jumpstart your dating lessons by telling you how to avoid the bad readers and how to separate the marketing gloss from the real world. The insight you gain in this chapter, along with real-world examples, helps you understand how to make a buying decision, what to look for, and how to test readers in your own environment.

Choosing a Hand-held, Mobile, or Fixed-location Reader

You can choose from three types of readers for your RFID network: hand-held, mobile, or fixed-location. Many criteria enter into this decision, including operating processes, system cost, read range, read rate, volume to be covered, and allowable physical locations of antennas. The operating process, required read data, and cost of installation will strongly influence the type of reader installation that you choose.

- ✔ **Hand-held:** Hand-held readers are acceptable if only one or two tags per read location need to be read at a time or if the volume is prohibitively large for fixed location readers. Using hand-held readers is similar to using a bar code reader, but unlike bar code readers, hand-held RFID readers don't require the tag to be clearly visible. Unfortunately, no hand-held readers with a high power output and long battery life are currently available. The advantage of a hand-held is that you can clearly scan one case or pallet at a time without interference from pallets close by.

- ✔ **Mobile:** A mobile reader located on a trolley or powered cart can be moved throughout a facility to read all the facility's contents economically. Hand-helds can augment this process for hard-to-read objects. The benefit of a mobile solution, such as the ACCU-SORT mobile tagging station, is that the printer, reader, and a bar code scanner can all be in an easily movable solution that communicates over 802.11 back to the main database. This one also has a middleware component incorporated and can help companies get compliant quickly with an investment of less than $100,000 in most instances, after proper physics testing and verification.

- ✔ **Fixed location:** Fixed-location readers may be located at all entry and exit points in a facility, on conveyors, at sort stations, or anywhere there is a *choke point* (a point where items must pass by). The benefit of a fixed RFID reader is its ability to automatically count and capture data without needing human involvement. I have set up fixed location readers that have achieved better than 98 percent success rate of reading items as they pass by on a conveyor. This type of automation and accuracy is what makes RFID the cat's meow.

After you have determined the type of reader to use, you're ready to evaluate it, as described in the next section.

Reading between the Lines: Critical Buying Criteria

Like a lot of people, I initially judged RFID readers by how far they could read. But before long, I began implementing real-world deployments for clients and found out what was really important. Because dock doors and conveyors aren't 50 feet wide, having a reader that could read at 25 feet isn't necessary. Dock doors are, at most, 12 feet wide, meaning that readers have to read only 6 feet (with antennas on both sides of the dock door) to cover most applications.

I clearly needed a new set of criteria for determining the difference between readers. I've distilled the more practical criteria for you in the following list. When you evaluate readers for your RFID system, follow these guidelines to determine the best choice for your business:

- **Determine all the costs involved:** Cost is always a major concern in business. In the case of RFID technology, more important than the purchase price of a reader are the hidden costs associated with installing and maintaining a reader *network* into the future.

- **Test reader performance:** As I point out in Chapter 9, various materials interact with radio waves in vastly different ways. Fluids, for example, absorb RF energy, whereas metals reflect it. Therefore, you need to choose an RFID reader that interacts effectively with your product material type(s) and environment. Just like an expensive FM radio may pick up stations that a cheaper radio cannot, so too do RFID readers work with varying degrees of success.

- **Assess connectivity:** With the state of technical specifications and software systems in constant flux, the ability to update, maintain, and monitor reader networks remotely is critical to long-term success. The bottom line when you look at an RFID network is that you consider the entire *ecosystem,* the many nodes of a highly distributed system. You need to consider four key factors relating to the connectivity of distributed reader networks and ease of deployment within an existing infrastructure. Those four key factors are

  - *Network connectivity:* How many ways can the reader send data to the outside world — over an Ethernet port, via RS-232, wirelessly, and so on.

  - *Manufacturer's configuration software:* What tools are added to the reader to allow management and configuration via the various connection methods? Sometimes, the commands are very different, depending on whether you're issuing them through a Web portal with a user-friendly interface or through the serial port connection with command line language.

- *Manufacturer's data export software:* Some readers can filter and smooth tag reads, whereas others will just export raw read data. This capability ties into your middleware architecture, which is explained in more detail in Chapter 10.

- *Upgradeability:* Check for features that help you upgrade. For example, can you use the same reader if you decide to expand interrogations zones and would rather talk to the main network wirelessly over 802.11? Will the firmware allow upgrades to new protocols? (I introduce the protocols in Chapter 2.)

Because complex physics underlie an RFID reader network, it's not very effective to simply attach the hardware to an existing network management system (NMS) or monitoring package. Many early adopters found this out the hard way: their NMS reported that the reader was alive, but it couldn't determine that the readers wasn't reading a single tag in the interrogation zone. You need middleware (see Chapter 10) and a monitoring system (see Chapter 14) that can handle RFID data.

✔ **Evaluate how well a reader can be fine-tuned:** RFID systems must be finely tuned to maximize the capability of reader systems with each product at each location. Many people experience ghost or phantom reads because their readers are improperly configured. If you're subject to one of the big mandates, your read rates must be 100 percent. After you get a system up and running, it becomes painfully clear that control over a large number of RF parameters is important. This ability to fine-tune will give you the control you need.

The sections that follow examine these four key guidelines in more detail.

Consider all the costs involved

When considering the reader costs for your RFID network, you need to take into account the costs associated with the purchase price, installation, and maintenance of readers.

Purchase price

The purchase price of your hardware is the foundation for evaluating a reader network's cost. Although the cost may seem straightforward, many reader manufacturers constantly revamp their pricing strategies, change their volume discounts, and offer marketing incentives for brand name clients. If you are a Fortune 500 client with brand-name recognition, put your negotiating hat on before talking to these hardware guys, or use an integrator with a lot of experience in buying large quantities of hardware.

Many reader manufacturers, particularly the newer venture-backed companies, are willing to literally give away readers to get a brand name client. In addition to securing the client, they hope the big-name client gets used to using its readers. Then, when the client rolls out beyond a one- or two-reader pilot, manufacturers hope the client *buys* a bunch of its readers. Like many of the dot.com companies of the '90s, reader manufacturers figure that if they give readers away at below cost, they will make it up in volume. That seems to be the case for the earliest of success stories — Matrics — which had the most marquee clients, including Wal-Mart, and which was purchased in 2004 by Symbol Technologies for $230 million.

Another interesting factor is the Federal Communications Commission's (FCC) requirement that readers be sold with approved antennas. The law states that at least one approved antenna must be sold with each reader. But reader manufacturers mostly resell off-the-shelf antennas at a significant mark-up. You can dramatically reduce the purchase price of RFID antennas by sourcing antennas through a major original equipment manufacturer (OEM) such as Cushcraft or Sensormatic. In some cases, you can cut antenna prices by as much as 50 percent.

Installation costs

RFID readers require access to 110 volt AC power or a router set up for Power over Ethernet (PoE). Bringing power to all the areas where you have readers set up may increase costs significantly, depending on how extensively the site is currently wired for power and network connectivity.

The reader's design (or form factor) comes into play in the installation as well. With one reader manufacturer, AWID, the reader and antenna are a single unit, eliminating messy coaxial cables that are somewhat sensitive to their environment. On the opposite end of the spectrum, Symbol uses fragile, quick-disconnect terminal connectors to tie its antenna to the reader via coax cable. Therefore with a reader like AWID, you can mount the reader anywhere there is power and connectivity. But if you need a reader and antenna combination in more areas, you need to purchase a multiplexer to add those additional antennas. The Symbol reader requires more protection for the cable connections and works best in a dedicated RFID rack. These little implementation issues add directly to cost when you consider an entire network of readers.

Maintenance costs

The on-going maintenance costs for RFID are still relatively unknown because of a lack of data from manufacturers. Here are some of the most common maintenance issues:

- ✔ **Replacing readers:** Reader failure often occurs because of conditions like heat, humidity, or extreme cold.
- ✔ **Upgrading for new protocols and standards**
- ✔ **Changing configurations**
- ✔ **Replacing damaged antennas**

All the major reader manufacturers provide some on-going technical support. For example, Alien requires attendance at its "RFID University" — with an associated price tag of $5,000 per person — before it offers tech support. ThingMagic relies on an outside party, Tyco, for support. If your maintenance issue requires on-site support, most vendors charge $150–$250 an hour and have a limited number of trained personnel.

You can minimize routine maintenance by taking these preventative measures:

✔ Place the reader — the most expensive and shock-sensitive component — away from high-traffic areas such as conveyors, shrink-wrap machines, or dock doors.

✔ Place antennas near high traffic areas in a specialized reader rack to avoid potential damage.

If you're forced to choose between putting either your reader or antennas in harm's way of forklifts, dock workers, or high-speed machinery, choose the antennas because they're relatively inexpensive ($100–300) compared with readers ($700–$3,000).

Test reader performance

The good news is that just about every reader manufactured can read the 6 feet or so required to read across a dock door. The bad news is that when you inject other variables like speed and material, things get a little more interesting.

Performing a basic distance test

To understand how each reader performs at various distances, you need to set up a basic distance test. The distance test is an accurate way to measure how distance affects a reader's performance. You can perform this test by following these steps:

1. **Find an empty warehouse or outdoor location with little or no ambient electronic noise (AEN).**

   To find out how to test for AEN, refer to Chapter 7.

   The ideal test site for reader performance is in an *anechoic chamber,* which is a closed room of 3, 5, or 10 meters that has no ambient electronic noise and does not reflect any RF waves. This is a pure test of how well the reader propagates a wave in a vacuum. However, RFID readers live and work in the real world, so you need to test them in the real world, where you can put the quality of their electronics and sensitivity of their receiver to the test.

2. **Find an empty cardboard box and affix to it an RFID tag that matches the protocol of the reader you want to test.**

   Class 1 tags always work better with Class 1–specific readers, and Class 0 tags always work better with Class 0 readers. For an introduction to protocols, see Chapter 2.

3. **Lay out a grid on the floor starting at the reader and moving away in 1-foot increments, marking on the ground every foot.**

4. **Set up the reader in the factory default setting (usually full power) with a single antenna attached.**

5. **Mount the antenna 3 feet off the ground on either a PVC mounting bracket (like the one shown in Chapter 8) or a camera tripod.**

6. **Put the cardboard box on the SKU (stock-keeping unit) testing table (also described in Chapter 8) or a similar nonferrous stand.**

7. **Measure the number of reads over a 60-second time period at 2, 4, 6, 8, 10, and 12 feet away from the antenna.**

8. **Record the results from each reader tested and compare the numbers.**

The top readers on the market today from a pure read range perspective are from Alien and Symbol (when used with their own tags). The top readers from a multi-protocol perspective are OMRON readers (when used with their own antennas) and SAMSys. The Tyco/ThingMagic reader also does well in a multi-protocol environment. But nearly all the major readers can read well at 6 feet, so the read range is likely to have little to do with your buying decision.

The reader software may skew your results because numerous parameters exist within each reader's firmware that are germane to each reader. Tuning these variables to represent identical settings is difficult. This disparity in firmware and firmware settings is one of the difficulties in measuring reader performance on an even playing field. What one manufacturer refers to as a *global scroll mode* might be called *setting 6* by another, and only intimate knowledge of many readers enables you to find commonalities. With all the major readers I've tested, I've found that every reader from *any* manufacturer can read effectively out to 6 or 7 feet in a clean environment, giving a maximum theoretical dock door width of 14 feet.

Setting up a conveyor test

One of the most common locations of RFID readers is on the conveyor line. Conveyors are everywhere — from production facilities, to distribution centers, to the back of stores. The constraint of having only a second or two to read an object as it passes through an interrogation zone adds to the complexity of designing the reader network.

The mandates from many major retailers and the Department of Defense (DoD) have specific speed conditions associated with them. Wal-Mart, for instance, requires that items be readable at 600 feet per minute (fpm) at a maximum of 6 inches of separation between each case. The easiest way to test this scenario is to set up a conveyor at 600 fpm. But unless you've seen something moving at 600 fpm, you probably don't have an appreciation of how fast that really is.

You can effectively test for performance at a certain speed in two ways: using a conveyor or simulating the speed of a conveyor with a Cyclotron.

- ✔ **The conveyor method:** Building a small conveyor that can reach that kind of pedal-to-the-metal velocity is difficult and very costly. I know one systems integrator who spent over $200,000 building a loop conveyor. The best way to work around the velocity and acceleration issue, while maintaining the purity of a conveyor system, is to build a *loop conveyor* that can gradually build up to that speed. In order to keep cases of your product from flying at all angles, you need to put rails up around the corners.

- ✔ **ODIN technologies Cyclotron:** A safer and easier way than building your own loop conveyor is to use a product called the Cyclotron. After trying to chop and trick out conveyor motors in order to exceed the 600 fpm mandate and over-engineer a system, the ODIN technologies team designed and built a device capable of accelerating cases of product up to 1,200 fpm (twice Wal-Mart's requirement) and called it the *ODIN technologies Cyclotron*. The Cyclotron safely accelerates cases of product and holds them at a constant speed indefinitely without fear of toppling cases or damaging the product. ODIN technologies is happy to share the design with end-users so they can build their own device.

The Cyclotron is essentially a giant centrifuge that has a 3-x-3-foot wooden or plastic pod at the end of an 8-foot axle made of wood or fiberglass. Rollers guide the axle as it spins around on a 4-x-4-foot tabletop. This enables you to complete accurate, safe, and quiet speed testing in a small area. A simple digital speedometer allows accurate calibration of the device down to the tenth of a foot per minute. The ODIN labs have both a conveyor and the Cylcotron, but the results and ease of use of the Cyclotron have the conveyor collecting dust. Figure 10-1 shows a line drawing of the Cyclotron.

Running a conveyor test

Here are the general guidelines for running a conveyor speed test:

- ✔ **Set up the readers in their fastest polling mode and read at gradually increasing speeds.** I like to test at 400, 600, and 1,200 fpm. You should employ high speeds to probe the upper threshold of reader performance.

- ✔ **Test a baseline RF-friendly product, such as a cardboard box.** For the baseline test, tag an empty cardboard box with the same type of tag used on each SKU and accelerate the box to the same speeds mentioned previously.

✔ **Test around your most RF-unfriendly items.** If you have items with high metallic or liquid content, test them on the Cyclotron or conveyor.

✔ **Test your most representative or popular product.**

✔ **Measure at least 50 passes.** You're looking for high-integrity data. To ensure that nothing unusual affects your test results, you need to complete the test enough times to make your data statistically significant.

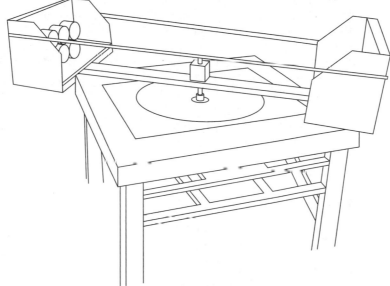

Figure 10-1:
The ODIN technologies, Cyclotron for testing performance at high speeds.

Conveyors come in all shapes and sizes. Rollers can be all metal, others made of composite materials, and some can have rubber sections of treading. Some tracks are just over 1 foot wide, whereas others can be 3 feet wide. The distance from one section of conveyor to another can vary by 10 or 15 feet. The electrical noise generated by each one can be vastly different. As you can see, *there's no one standard conveyor.* Therefore, don't try to compare results of conveyor testing from one conveyor to another, particularly if you're testing different readers with various configurations, power outputs, and antenna designs. Trying to talk to your peers or sister companies about conveyor testing can lead you down a confusing path.

Test conveyors in your live environment to verify your scientific static testing and use that testing to set a threshold for what level of performance you need to achieve when you perform you static testing or define your required tag performance index (TPI). For more information on the TPI, see the sidebar in Chapter 9. Beyond that, the results are widely variable with multiple dependencies — not something any pocket-protector-wearing scientist would want to hang his slide rule on, that's for sure.

Interpreting the results

The goal of performance testing is to compare each reader in a manner that eliminates all variables except those that you're examining. The distance test validates that the readers you're testing have some common baseline of performance, maybe 6 or 7 feet on RF-friendly material. With that information, your next challenge is to examine the conveyor test:

- ✔ Your **control variables** are the speed at which you're testing and the materials you're testing — an RF-friendly material (your baseline), your toughest product, and perhaps your most representative or popular product.

- ✔ The **dependent variable** is the average number of successful read cycles completed per pass because you're testing the ability of the reader and its electronics to very quickly interrogate tags.

You can accomplish experimental equivalency by dividing each reader's raw data by its own unique *baseline* data to normalize the SKU data generated by each reader. For example, reader *x* might successfully read bottled water three times per pass but read an empty cardboard box (the baseline) six times per pass. This would result in a rating of 0.5 for your bottled water testing (3 ÷ 6). This normalization is the only way to easily eliminate all firmware- and software-related variables (for example, polling cycles and frequencies) and compare each reader's performance against its own unique baseline.

Make sure the same cases of product are tagged with the *exact same* tags for all materials and readers involved, effectively eliminating case and tag variance as well.

ODIN technologies labs tested the performance of four EPC-compliant readers, and the results are shown in Figure 10-2. This figure clearly illustrates the role of speed in readability. As speed increases (the y-axis), each data group becomes shorter, indicating fewer reads per pass and diminished performance. It's interesting to note that both Alien and AWID exceed the 1.0 threshold at 400 fpm, indicating that these readers performed better communicating with tags attached to paper products than they did in the baseline case. Taking the data from the tests and presenting it in a simple graphical manner, as shown in Figure 10-2, gives you an easy-to-read side-by-side comparison.

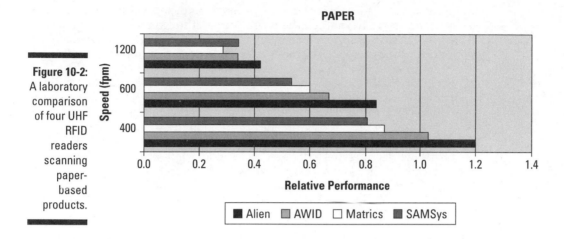

Figure 10-2:
A laboratory comparison of four UHF RFID readers scanning paper-based products.

Figure 10-3 shows the same four readers, but this time the object being tested is a metallic case. The change to a metallic SKU has a clear impact on the results: Performance is negatively impacted for all manufacturers at all speeds. Another significant finding is that Symbol offers superior performance when metallic objects are involved, with Alien now ranked third. Again, the role of speed is clear: Performance diminishes with increasing speed. The specific results for your products will vary, but the testing protocol and presentation will prove a useful factor in determining reader performance.

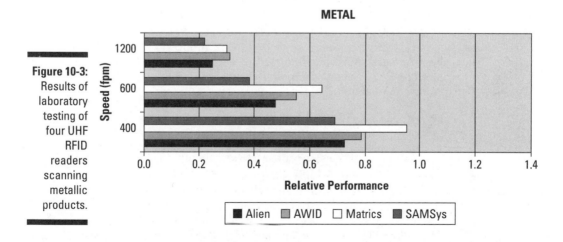

Figure 10-3:
Results of laboratory testing of four UHF RFID readers scanning metallic products.

To get the best results, you have to figure out how to make the readers perform as close to each other as possible. This is more difficult than you might think. For instance, the Symbol reader has only one basic acquisition mode, which is analogous to Alien's "Inventory" command, suggesting that Alien has programmed more flexibility into its firmware. Although the inventory mode works well in situations in which many tags are in the field simultaneously (like a dock door or shelf), it is not preferred in conveyor applications in which only one or two tags are in the field for a short period of time. The likely reason for this limitation is that Symbol's latest reader (AR-400) builds on its predecessor (SR-400), which was designed exclusively for Class 0 tag negotiation. The result of this limitation is a reduction in the total number of reads per pass relative to other readers. Although the speed test protocol cancels the effect of this issue by dividing reads per pass by each reader's own baseline, keeping these differences in mind is important. That's because the programmability and software is providing a perceived difference in performance that needs to be normalized as accurately as possible.

Assess connectivity

An RFID reader is only as good as the data that it can send out. Therefore the connectivity — how information gets from the reader to upstream applications is critical to the performance of your RFID network.

Choosing a connectivity type

While not exactly like deciding between blondes, brunettes, and redheads, RS-232, RS-485, and Ethernet are still pretty sexy choices. When deploying an RFID system, the choice of network connectivity for the RFID reader devices is an important consideration.

Historically, RFID readers have tended to use serial communications for their connectivity, with an equal reliance on RS-232 and RS-485. These days, most manufacturers gravitate toward Ethernet, with several working toward or implementing Power over Ethernet (PoE) and even 802.11 wireless connectivity. Your connectivity choices are as follows:

- ✔ **RS-232:** The RS-232 protocol provides a well-known and reliable system for short-range wired communications, but its limitations are rife:

  - Communications speeds are low, ranging from 9600 bits per second (bps) to 115.2 kilobits per second (Kbps) — think of slow modems from way back in the '80s.

  - Cable length is restricted to 30 meters.

  - There is no error-checking or error control.

  - It is a point-to-point communications system, requiring individual cables to be installed between each network device and the control system. Because of this, it requires extra cabling, which drives up the cost of deployment.

✔ **RS-485:** The RS-485 protocol is an improvement over RS-232 in the following ways:

- It allows significantly greater cable lengths (up to 1,200 meters).

- It handles higher speeds (up to 2.5 megabits per second [Mbps]).

- As a bus protocol, it allows multiple devices to be connected to the same cable.

However, it requires special cabling, which can drive up the cost of an RFID installation. But if you're comparing RS-232 and RS-485, RS-485 wins hands down.

✔ **Ethernet:** Since it has been available, Ethernet has been the connectivity solution of choice for most RFID installers for many reasons:

- Many integration locations already have Ethernet infrastructures in place, and its ubiquity ensures simpler installation and lower integration costs.

- Its speed is more than sufficient for individual RFID readers.

- The reliability of the TCP/IP protocol over Ethernet ensures the integrity of the data transferred.

The only shortcoming is that TCP/IP packets can travel only 100 meters down a single CAT-5 cable.

Many manufacturers recommend the use of protocol translation (PT) devices to convert from RS-232 and RS-485 to TCP/IP. This combines the lower development costs associated with serial communications with the higher speeds and reliability of Ethernet. RFID companies have traditionally used products from a company called Moxa for this purpose, but any RS-232-to-TCP/IP converter should work, and they generally cost under $100 each.

Many of today's new readers have Ethernet connectivity. Symbol, Alien, AWID, and SAMSys all have Ethernet connectivity in their latest readers. The best connectivity arrangements in a reader provide RS-232 and RS-485 for backward compatibility with previous models, while also containing both a standard RJ-45 Ethernet port and a PCMCIA slot (like the one on the side of your laptop that you can plug other devices like a wireless card into). The PCMCIA slot can house an 802.11 Wi-Fi PCMCIA card. Figure 10-4 shows some of the connectors and indicators that you might find on a new reader.

Although most previous-generation readers used serial communications for their network connectivity, the current generation of readers provides access to their systems over Ethernet. This is a significant advantage over serial communications because serial line communications provide little flexibility and slow communication speeds. Ethernet connectivity allows simpler installation and lower rollout costs, combined with greater ease of management because all Ethernet-connected network units can be remotely monitored to some degree.

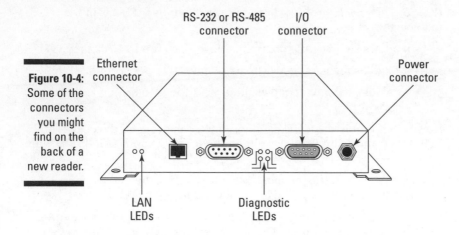

RS-232 or RS-485
connector

I/O
connector

Ethernet
connector

Power
connector

Figure 10-4:
Some of the
connectors
you might
find on the
back of a
new reader.

LAN
LEDs

Diagnostic
LEDs

Reader manufacturers supply antennas and 15–25-foot cables. Before you purchase your reader and antennas, make sure you know exactly where each one will be mounted so you can order cables of the proper length. Likewise, you need to find out from the manufacturer how long the power cable is so you can account for AC power if you use a standard outlet to power the reader.

Using the software for device configuration

Several manufacturers provide their own software, which interfaces with their equipment. Although most of the provided software is not useful for production deployments, it is useful for performing basic device configuration and for providing a demonstration of the unit's capabilities. If you'll be configuring and deploying the readers yourself, understanding the complexity of this on-board software is important. Many manufacturers, like Alien and SAMSys, require you to use command-line level controls for changing the configuration. Other manufacturers, like Symbol and OMRON, have a very simple Web interface that is easy for novice users, but at times not detailed enough for experienced deployment teams.

For fully integrated deployments, you need to install RFID reader control software systems, such as Sun Microsystems' EPC Information Server, ConnecTerra's RFTagAware, OATSystems' Senseware, or GlobeRanger's iMotion suite. These systems provide a consistent interface to many types of readers, allow tags to be written to (commissioned), and eliminate the need to write custom software to communicate with each type of reader. Keep in mind that these demonstration programs are not production quality; most of them are fragile, and system crashes are commonplace.

Some reader manufacturers provide a hidden supervisory mode in the demonstration software that gives experienced users access to advanced features. This mode is not documented or supported by the manufacturers because it makes it possible for inexperienced users to cause physical harm to the system. However, if you or someone on your team is an experienced software hack, this is where you can get to some really powerful commands.

Reducing cabling needs to save money

When implementing RFID solutions, all but one or two of the current-generation RFID readers require separate electrical and network connectivity, and in the case of RS-232 or RS-485 devices, custom cabling is necessary for connectivity. As you might have guessed, all these specialized cables that need to be routed around a location can increase the cost of implementation significantly.

There are two possible solutions for reducing cabling requirements and therefore costs; both solutions require Ethernet connectivity:

- **Wireless connectivity:** IEEE 802.11 Wi-Fi connectivity eliminates the need for a separate network connection but requires a local electrical supply to power the units.

- **Power over Ethernet:** The IEEE 802.3af Power over Ethernet (PoE) standard permits power and network connectivity to be delivered over the same physical CAT-5 cable but requires an upgrade of network gear to a specialized PoE switch.

Neither solution is a panacea for connectivity and cabling requirements. PoE is restricted to a maximum power output of 15.4 watts, a power level that can be significantly reduced as it travels along lengthy Ethernet cables. This power limitation requires more modest RFID tag reader power consumption to be feasible. The good news, however, is that expensive electrical connections will not have to be run at each reader station with PoE. Similarly, wireless connectivity can also prove problematic: It requires a Wi-Fi infrastructure to be installed, and the integration of Wi-Fi units into each RFID reader can drive up the cost of each unit. But on the other hand, the cost of bringing CAT-5 cable to each reader location is eliminated, so once again, it's a trade-off.

Real-world communications networks exist in most modern warehouses and vary in type from wireless to CAT-3 (serial) to CAT-5 (Ethernet), with almost all new installations using CAT-5. All readers will eventually support an Ethernet network, and it is heartening to see that many manufacturers already support this standard. If you like a particular reader that doesn't support Ethernet, ask the manufacturer whether Ethernet support is in the development timeline.

Evaluate how well the reader can be fine-tuned

When evaluating a reader for tuning capabilities, look for the following features:

- ✔ Forward compatibility with future tag protocols
- ✔ Tunability
- ✔ Flexible power output
- ✔ Clean RF output
- ✔ Antenna programmability

Forward compatibility with future tag protocols

Because the RFID industry is still developing, protocols are in flux — from EPC standards to ISO standards. Thus, choosing an *agile* (multiprotocol and multifrequency) reader to protect your investment against future changes to standards and to avoid technology obsolescence is vital.

Many reader manufacturers claim forward compatibility with the newest generations of tag protocols, but only one or two readers have shown the ability to read all the tags approved by Wal-Mart and the DoD (Class 0, 0+, and 1) and to be forward compatible with the new Generation 2.0 protocol. Symbol and SAMSys have readers for both Class 0 and Class 1 protocols and have proven compatibility since mid-2004. ThingMagic, Tyco, and OMRON have all claimed compatibility, and I expect those readers to be compatible based on their claims.

When you purchase readers, make sure you get a guarantee *in writing* that says the readers will be forward compatible with emerging standards. If the reader ends up not being easily upgradeable via firmware, you want to make sure that you get the reader manufacturer to pay for the cost of putting new readers in. Most top-tier reader manufacturers are willing to make this guarantee and stand by their equipment.

Because most of these readers have some sort of a digital signal processor (DSP) chip, which I cover in Chapter 5, there is reason to believe that all readers *could* be software upgradeable. Find out if your chosen reader has a standard DSP from a popular chip manufacturer like Texas Instruments. These manufacturers are more likely to offer upgradeable chips and support, so this information can help you gauge the likelihood of the firmware being forward compatible.

Tunability

When choosing a reader, you need to look at what parameters it enables you to tune. The process of extracting maximum performance from any reading device requires tunability that goes far beyond the functionality provided by demonstration software. A host of parameters must be tuned to optimize reader performance, including acquisition cycle design (which relates to the timing and number of wakeup, read, and sleep commands included in each acquisition cycle), acquisition frequency, and timing. For best performance, each of these reader parameters should be configured for each unique SKU and for the specific operating environment. However, you have to tune these parameters to the common denominator.

Some manufacturers preset these parameters and package them in a single command for use in their demonstration software. Alien, for example, provides the ability to quickly capture data from any tag in the field through the Global Scroll acquisition command. This mode is preferred in a fast-moving environment such as a conveyor application, in which no more than one or two tags enter the field at a given time. The Global Scroll command allows for very fast interrogation and response, so it can collect a small amount of information very quickly. Conversely, the Inventory mode provides the ability to negotiate hundreds of tags simultaneously by using an algorithm that efficiently steps through all EPC numbers in the field. Though much slower, Inventory mode makes reading all cases on a pallet possible by solving the problem of data collision.

Both SAMSys and Matrics have systems that create custom configurations based on polling rates, protocol choices, and other factors. Matrics creates separate read-point classes. The classes are a configuration of each antenna attached to an individual reader. Configuring the Matrics reader allows you to set the following options:

- ✔ **Antenna type** allows you to choose from five different preset antenna configurations, based on your specific usage need — AREA (long range), SHELFv1 (existing short range shelf type), SHELFv2 (next-generation short range shelf type), COMBINED1, or COMBINED2.

- ✔ **Scan Period** dictates how often the read point is to be checked for tags.

- ✔ **Retry** tells the reader how many times to repeat the read command each time a scan is to be performed.

- ✔ **Air interface** — Read All, Class 0, or Class 1 — determines which protocol the reader should use.

- ✔ **Gain** designates (as a percentage) the antenna's power setting.

The custom configuration options are more complex to understand and use in an operational environment. However, they could result in a higher read rate and great success after some experience with the settings. This is why having your own testing laboratory or environment (covered in Chapter 8) is an important part of understanding the performance of each reader.

Reader power and path loss

Understanding some of the parameters of a typical UHF reader will help you further appreciate the proper reader network architecture. With an understanding of some of the basic math behind the physics, as well as some of the FCC guidelines that regulate the use of RFID, you can properly tune your readers for optimal performance. The maximum allowable power output at each antenna on a reader in the United States is 1 watt. This equates to

✔ Reader transmit power (P_r) of 30 dBm (an adjustable variable in most readers and a measurable one as well)

✔ Reader receiver sensitivity (S_r) is equal to −80 dBm, or 10^{-11} watts

✔ Reader antenna gain (G_r) is equal to 6 dBi

This is the information half of what you need to know, the second half is the tag component:

✔ The tag has a power requirement of −10 dBm, or 100 microwatts.

✔ The tag's antenna has a gain (G_t) of only 1 dBi.

✔ The tag's antenna has a backscatter efficiency (E_t) of −20 dB.

Understanding these six components, coupled with (pun intended) the known wavelength (λ) of 915 MHz of 33 cm (about a foot), you can determine the path loss of an RF field. Knowing the path loss can help you optimize your read zones. That's why I go through path loss contour mapping in Chapter 7. The following figure shows how the tag power (P_t) is calculated in a path loss. Knowing the tag power requirements (100 microwatts) can help you calculate the reader power (P_r) to optimize the distance of the interrogation zone. This is critical in preventing ghost or phantom reads.

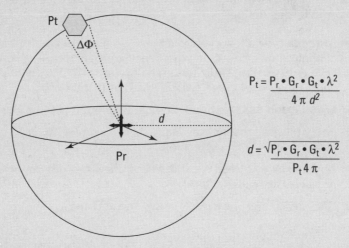

$$P_t = \frac{P_r \bullet G_r \bullet G_t \bullet \lambda^2}{4\,\pi\,d^2}$$

$$d = \sqrt{\frac{P_r \bullet G_r \bullet G_t \bullet \lambda^2}{P_t\,4\,\pi}}$$

Flexible power output

"Supersize it" is the general mantra of everything in the United States — the bigger, the better. As a guy who's owned two Hummer H-1s, I feel the urge to supersize on a regular basis. But in RFID, more is not necessarily better. The experience from ODIN's lab unequivocally — though counterintuitively — indicates that many tagged objects have a maximum performance threshold well below the maximum power output of a reader (1 watt at the antenna). In other words, you get better results trying to read things with less than full power — kind of like going faster in second gear. This is due to the fact that RFID tag backscatter can be drowned out by high-power transmissions on the receiving side. So having a reader with flexible power output is critical to achieving maximum performance.

Minimizing power output for each application to avoid unnecessary RF levels in a facility where other readers are bound to be operating is also helpful because you want to minimize the amount of RF noise in your environment so that readers don't interfere with each other. Proper power tuning is critical for an RFID network and should be given the utmost consideration when deploying a system. As reader technology improves, so too will control over variables like power.

Many reader manufacturers have a confusing power adjustment setting such as "Power: 1–10," with little or no explanation of what 1 is or how many decibels each level accounts for. Using a spectrum analyzer and power meter can be a valuable way of comparing relative signal strength. The small hand-held units, such as the Rohde and Swartz portable hand-held spectrum analyzer and an OMRON reader, are great for mobility in the field. As users become more sophisticated in their understanding the tuning capabilities, the readers will grow in sophistication.

Many issues attributed to "immature technology" are easily solved when applying the laws of physics throughout the system design and installation. Proper tuning prevents RF spillover from affecting adjacent systems, and as the "Reader power and path loss" sidebar shows, mathematical formulas and simple physics equations can help you understand this complex behavior.

Clean RF output

Along with the low-cost status of an EPC-compliant tag comes a simple timing device used to synchronize with the carrier signal emitted by a reader. If this signal is not constant phase (that is, if there are unwanted variances of the RF wave because of poor quality equipment), the tag clock will fail, making it difficult to establish communication with the reader. Therefore it is important that the RF emissions be as clean as possible.

More importantly, some RF signals generate *splatter*. Splatter occurs when the RF power of a reader transmitter causes undesirable frequencies to be generated. This "dirty" signal can cause interference with equipment operating at frequencies outside the RFID ISM band and break FCC compliance. Generally, the quality of RF output is attributable to the quality of components used to produce the reader.

Many readers are not being made in production volumes and, consequently, have poorer than normal quality control. You want to ensure any reader you test doesn't produce splatter and is in compliance with FCC regulations. Certifying all readers in the lab to make sure the reader and antenna combination complies with FCC rules Part 15 is critical to keeping your system legal and safe. The best way to ensure compliance is to set up the reader and antenna combination in the lab as it will be configured in the operating environment and then test it. Follow these steps to test for FCC compliance:

1. **Make sure the power setting is at the maximum allowable amount (which should be 1 watt).**

2. **Set up your spectrum analyzer with a center frequency of 915 MHz and a span of 50 MHz so that you can record all activity from 865 to 965 MHz.**

3. **Monitor the reader for 60 minutes and keep the hold lock ON in your spectrum analyzer to record all the read points.**

4. **Look for any signals outside the 902–928 MHz range.**

   If you have any signals outside the 902–928 MHz band, you've got a non-compliant system that needs to be sent back to the manufacturer.

Antenna programmability

In evaluating a reader, you also need to consider the programming capabilities of its antennas so that you can control how an antenna array works. The antenna acts as both the ears and the mouth of a reader — it transmits (talks) a signal to tags and then receives (listens) for the response. Antennas are critical to the RFID network and need to be carefully configured and set up. The antennas of multiport readers go through a switching sequence called *multiplexing.* No two antennas are ever active at the same time to prevent interference and data collision. Some readers make it possible to program this sequence, and others do not.

The advantage of sequence programmability is illustrated clearly in a portal application in which cases stacked on a pallet must be read. Pallets of some SKUs possess regions that are difficult to penetrate with RF radiation; tags in these regions require additional time to power up and communicate with the reader. As you develop your RFID program, conduct SKU testing that seeks to discover these regions. You must then be able to focus RF power on those regions and program the antenna sequencing pattern accordingly. So rather than using the default mode that activates the antennas in the following sequence:

   1, 2, 3, 4, 1, 2, 3, . . .

It may become necessary to adjust the sequence to penetrate a difficult region adjacent to antenna 3 as follows:

1, 2, 3, 3, 3, 4, 1, . . .

Always design the system for the most challenging use case.

Many readers build on this concept by making it possible to configure each antenna individually. Matrics offers configuration parameters such as antenna gain, the number of retries, the air protocol to use, and others. Alien takes antenna control one step further by providing antenna duration control, making it possible to adjust the amount of time one antenna is active relative to another. These custom tuning capabilities will continue to evolve in readers as people define the needs of their RFID network based on real-world experience.

Installing a Reader and Antennas

After you select your reader and antennas (which you will probably buy from a reseller or manufacturer), it's time to let the fun begin. I suggest installing the antennas and firing up the reader for the first time in a lab setting so you have more control and less activity around while you try to figure out these new pieces of equipment. You should also have a spectrum analyzer and network analyzer available for fine-tuning and verification. (See Chapter 12 for more on analyzers and the lab set-up.) The four basic steps to setting up your reader and antennas are as follows:

1. **Mount the reader.**
2. **Mount and connect the antennas.**
3. **Power up the reader.**
4. **Test the interrogation zone for RF path loss.**

These steps are described in detail in the sections that follow.

One of the challenges of setting up the RFID infrastructure is mounting the readers, antennas, and cabling in a warehouse or production environment. A simple solution for this is a specially designed RFID rack. Several manufacturers sell all-in-one RFID racks that hold the reader, antennas, cabling, indicator lights, and in some instances, an uninterruptible power supply (UPS). These units protect the reader and antennas and allow you to build in certain features that are important to your business process. Figure 10-5 shows the inside and outside of a ruggedized RFID rack. The figure on the left shows the interior with the antenna mounting bar and the shelves for readers, strobe lights, and the bottom compartment for a UPS. The figure on the right shows the rack with the protective cover in place.

Figure 10-5:
A rugged-
ized RFID
rack.

Mount the reader

Most readers have holes on the edges of the outside casing or a soft material you can drill directly through for mounting the readers to the desired surface. If you use an RFID rack, you don't have to worry about the hassle of finding a safe place to mount the units. Mount the reader by screwing it directly to the wall or plate that you fashion to hold it, making sure you screw it down in all four corners.

When looking for a place to mount the reader, keep the following points in mind:

- ✔ **Make sure that the reader is as far from harm's way as possible.** Mounting the antenna above a dock door, for instance, keeps it in the area of least activity; the downside is that it's tough to get to if you need to work on it.

- ✔ **Mount the reader in as clean and dry a place and in as normal a temperature as possible.** Any extremes in heat, humidity, direct sunlight, and so on have an adverse effect on the reader.

- ✔ **When you mount the reader, make sure you allow at least six inches of clearance all the way around the reader** for ample air flow to cool the unit and to allow cables to be easily connected.

In addition to understanding the sources of ambient electromagnetic noise (AEN), gleaned from completion of the Full Faraday Cycle Analysis detailed in Chapter 7, you also need to consider more direct and immediate electromagnetic interference on the reader (as opposed to on the RF waves emanating from the reader). Mounting the reader near generators, pumps, converters, electronic scales, AC switching relays, computer terminals, and the like may adversely affect the reader's performance.

Don't power up the reader yet. Powering some readers up without antennas attached can overload the circuitry and literally fry the electronics of the reader. Most new readers have circuit protection built in, but I've fried a couple of the old ThingMagic readers in my time by carelessly disconnecting the antennas before I powered the reader down. Thankfully, better reader design keeps this from happening in most instances, but making sure you keep antennas plugged in when the power is on is a good habit to get into.

Mount and connect the antennas

After you decide where you want your antennas and use the RF path loss contour map (see Chapter 7) to lay out the design, you can mount the antennas. Antennas generally come either with a material around the edges that's designed to be drilled through or with special mounting brackets attached to the back. You can mount antennas on anything from specially designed ruggedized RFID racks to custom-welded angle iron to lolly columns that hold up the ceiling. Remember that you want to protect the antennas from people or machinery; they can't be bumped, bashed, banged, or bothered with.

After the antennas are mounted around your interrogation zone, attach them to the reader in *sequential order*. Many people ignore this important point and just start screwing coax cables willy-nilly into random ports. Depending on the type of reader you choose, you have either eight or four ports. If you have eight ports, you most likely have both a transmit (Tx) and a receive (Rx) port, and your antennas have two cables coming from each of them. If you have four ports on the reader, the Tx and Rx ports are probably combined. To connect them in sequential order, first connect antenna 1 to reader connectors Tx1 and Rx1, then antenna 2 to reader connectors Tx2 and Rx2, antenna 3 to the Tx3/Rx3 connectors, and antenna 4 to the Tx4/Rx4 connectors last. Follow these steps to make the connections:

1. **Attach the large ends of your antenna connector cables to the large connectors on the antenna.**

2. **Attach the small ends of the cables to the corresponding connectors on the reader (antenna 1 to reader connectors Tx1 and Rx1, and so on).**

Power up the reader

Now the fun part: Fire up the reader and start reading. Powering up readers is pretty much the same as booting up a computer. After the reader powers up, it goes through a series of self-checks (you see the lights flashing on the outside of the reader while it does this). Make sure you have the right power supply. Connect the power supply to the reader port for power and then plug the power supply into a 24-volt DC power outlet. The reader's power light should come on. Depending on the reader, it may take a few seconds to a few minutes to boot up properly, but as soon as it does, you can start reading.

If you've done everything correctly and the power light is on, you can put a tag in front of an antenna and get a visual verification by seeing a transmit/receive or read signal. Here are the detailed steps:

1. **Hold the tag directly in front of and at least 1 foot away from the antenna.**

   At this point, you don't want to move the tag around too much because the reader may not be optimally configured. Just put the tag in front of the antenna and look for the read light to flash on.

2. **Check the validity of that signal light on the reader by connecting the reader up to your laptop via either the RS-485 port on most readers or the Ethernet port. You can then start up the simple demo program that comes with all readers to verify that the reader is actually generating good data.**

3. **When you're satisfied that the system is working properly, organize all the cables and permanently mount the cables with wire ties and mounting brackets so they are neatly out of the way.**

   Running the cables along mounting brackets or enclosing them in protective piping is always a good idea.

Test the interrogation zone for RF path loss

Using the technique described in Chapter 7, set up the spectrum analyzer and verify the strength of the RF field in your interrogation zone. By putting your testing equipment to work after the installation, you can determine the overall area of your interrogation zone and configure it to read only where you want it to. Knowing that a tag generally requires –10 dBm or 100 microwatts to power it up tells you where the outside edge of your interrogation zone will be.

After you're up and running in a stable state, try to connect to the reader via its Ethernet connection and begin to investigate the other options that might be available through the configuration GUI (graphical user interface). The reader's GUI often gives you access to different configuration options and usually an intuitive interface for making and saving changes. Almost all the top readers today (Alien, Matrics/Symbol, OMRON, SAMSys) have built very intuitive GUIs.

Chapter 11

Middle Where? It's Not Just about the Readers

*A*ll you baseball fans out there probably know that the greatest baseball team of all time, the 2004 Boston Red Sox, would not have been able to pull off the best comeback in baseball history (down three games to none in a best of seven series with the Yankees, they swept the National League Champion Cardinals, and became the only team in baseball history to win eight games in a row culminating in a world championship) with only their starting pitchers. The key to their victory was having a great starting rotation, a good closer, and strong middle relief pitchers. The exact same scenario rings true in the RFID world. If you want to be the world champion of RFID deployments, you have to look at the physics and infrastructure covered in this book and have an end application that can turn the data into useful intelligence. But you won't be anywhere without the right middleware to connect the infrastructure with the application.

Tag and reader physics solves only the problem of being able to *capture* RFID data. To uncover the operational benefits of RFID — like reducing out-of-stock situations or decreasing labor requirements in the receiving process — you must *process* the incoming RFID data and intelligently *integrate* it into your business applications. This is trickier than it might sound because you can't just link existing software applications to RFID readers and hope for the best. Why?

- **First, not all of the incoming RFID data is valuable.** Duplicate reads and excess information must be filtered out so as not to bog down the network and end up with confusing information inside your applications.

- **Second, not all readers speak the same language.** Building custom integration logic for each brand of reader will quickly eat up your RFID deployment team's time and budget.

- **Lastly, different RFID information needs to be passed off to different applications and data stores.** For example, reads at the loading dock may need to be passed off only to the plant's local warehouse management application, while demand stream information coming from a retailer may need to be sent all the way up to the supplier's enterprise demand forecasting solution.

These capabilities are the heart and soul of a new breed of software called RFID middleware, which Forrester Research defines as "platforms for managing RFID data and routing it between tag readers or other auto identification devices and enterprise systems."

This chapter takes you through the specific features and functions that RFID middleware provides, as well as the various types of offerings that are currently on the market. I help you understand the different standards in the RFID middleware world and show how to plan and implement the RFID middleware solution that is right for your situation and current technical environment.

Filter, Smooth, Route: Understanding What You Need Middleware to Do

The scope of what RFID middleware needs to do varies depending on whom you talk to, but I walk you through the full list of functionality that a complete end-to-end middleware solution should include. Many early RFID middleware solutions focused on features like reader integration and coordination, electronic product code (EPC) track-and-trace tools, and baseline filtering capabilities. But these are just a subset of the many features that complete RFID middleware platforms must provide.

To stand the test of time as a complete solution, RFID middleware must include a balanced combination of seven core capabilities. These capabilities — starting from connectivity and moving up the stack — are as follows:

- **Reader and device management:** RFID middleware needs to allow users to configure, deploy, and issue commands directly to readers through a common interface. For example, users should be able to tell a reader when to "turn off" if needed. In some instances, middleware vendors offer plug-and-play-like capabilities that let deployment teams dynamically sense a reader's presence and link to it without having to write any code. Integration with other Auto-ID technologies, like sensors and biometrics, is also important.

✔ **Data management:** After RFID middleware captures EPC data from readers, it must be able to intelligently filter and route the data to the appropriate destinations. Look for middleware that includes both low-level logic (like filtering out duplicate reads) and more complex algorithms (like content-based routing). Comprehensive solutions also offer tools for aggregating and managing EPC data in either a federated (multiple repositories of data) or central data source.

✔ **Application integration:** RFID middleware solutions need to provide the messaging, routing, and connectivity features required to reliably integrate RFID data into existing SCM, ERP, WMS, or CRM systems — ideally through a services-oriented architecture.

In case you're not familiar with these terms, that's *supply-chain management* (SCM), *enterprise resource planning* (ERP), *warehouse management systems* (WMS), and *customer relationship management* (CRM) systems. A services-oriented architecture is essentially a collection of services. These services communicate with each other. The communication can involve either simple data exchange or two or more services coordinating some activity, such as order placement or inventory control.

Middleware also needs to provide a library of adapters to popular WMS and SCM applications like SAP or Manhattan Associates, as well as application programming interfaces (APIs) and adapters for using standard technologies like JMS, XML, and SOAP to integrate with other third-party applications.

✔ **Partner integration:** Some of the most promising benefits of RFID will come from sharing RFID data with partners to improve collaborative processes like demand forecasting and vendor-managed inventory. This means that RFID middleware must provide B2B (business-to-business) integration features like partner profile management, support for B2B transport protocols, and integration with a partner's data over communications such as EDI, Web-based systems like AS2, or eventually a well-engineered system specifically for EPC data.

✔ **Process management and application development:** Instead of just routing RFID data to business applications, sophisticated RFID middleware platforms actually orchestrate RFID-related end-to-end processes that touch multiple applications and/or enterprises. Using inventory replenishment as an example, if your system understands that you have a certain amount of one item coming through the door and the receiving process in the back of the store is tied to the point-of-sale data, you can accurately know when the inventory level becomes critically low and send a machine-generated message to the distribution center to order more product, without needing human involvement.

Key process management and composite application development features include workflow, role management, process automation, and UI (user-interface) development tools. These specific tools help you create solutions that fit in with your existing applications and reap the benefit of machine-to-machine communication in a custom environment.

✔ **Packaged RFID content:** RFID middleware platforms that include packaged routing logic, product data schemas, and integration with typical RFID-related applications and processes like shipping, receiving, and asset tracking are major assets. Why? No one wants to start from a blank sheet of paper, and this content can give you a head start on your RFID projects.

✔ **Architecture scalability and administration:** There's no question that RFID adoption is going to produce a lot of data, and RFID middleware is the first line of defense for reliably processing that data. This means that RFID middleware platforms must include features for dynamically balancing processing loads across multiple servers and automatically rerouting data if (or should I say *when*) a server fails. These features need to span all tiers of the architecture — even the devices that are located near or on the actual readers. I explain the key tiers of a network a little later in this chapter in "Piecing Together a Middleware Architecture."

Exploring Middleware Vendors and Their Offerings

It's no surprise that vendors are flocking to the RFID middleware market, which is riddled with mandate-driven user companies whose RFID budgets are increasing. But as is typical of most emerging markets, the vendor landscape is far from simple. There are small start-ups with unique solutions, big companies with solutions only on paper (often called *slide-ware* or *vapor ware*), and everything in between. RFID middleware vendors are crawling out of the woodwork, although these vendors generally fall into one of five key segments. To get the full range of functionality from your middleware (all the functions I listed in the preceding section), you likely need to choose middleware from multiple vendors and segments. Among the five segments, each brings different expertise to the table:

✔ **Reader vendors:** You may be asking yourself, why don't the readers themselves provide these middleware features? Some of them do, and I expect many more to do so in the future. But the middleware features that readers provide are very basic and typically limited to things like stripping out duplicate reads. To do more sophisticated filtering and routing, you need more contextual information, like data from multiple readers and business logic that may reside in existing business applications. This type of information is not available to individual readers that are located on plant floors or in retail outlets.

That's no ordinary idiot — that's a Savant

The crazy standards and terminology of RFID can confuse even the most educated technical gurus and RFID veterans. As you may know, the Auto-ID Center at MIT was the breeding ground of many of the early RFID standards. One of those standards was called the *Savant,* which the Auto-ID Center originally defined as "a data router that performs operations such as data capturing, data monitoring, and data transmission." The Auto-ID Center envisioned these Savants working together in a hierarchical framework to manage EPC data throughout the enterprise. Sounds pretty similar to the definition of RFID middleware that I just laid out, right? It is. Much of the work that the Auto-ID Center did to architect and define the Savant standard is the basis for what we now consider to be RFID middleware.

When the Auto-ID Center closed in October of 2003, EPCglobal, Inc., a joint venture between EAN International and the Uniform Code Council, Inc., carried forth the research completed by the Auto-ID Center. The previous work on the Savant standard became a significant input to efforts now underway and was spearheaded by EPCglobal's Software Action Group.

As of February 2005, EPCglobal's Software Action Group is currently working to define standards for several RFID middleware functions, including

✔ Capturing, securing, and accessing EPC-related data.

✔ Obtaining filtered, aggregated data from several sources. (This standard is sometimes referred to as the ALE standard.)

✔ Exchanging data and commands between hosts and readers to do things like read tags, write to tags, and kill tags.

✔ Configuring, provisioning, and monitoring individual readers.

EPCglobal doesn't actively use the Savant terminology anymore. But because EPCglobal has not officially published new standards for all these functions, you still frequently hear the term *Savant* used interchangeably with *middleware.* You can find the latest information about middleware standards and the Software Action Group on the EPCglobal Web site, www.epcglobalinc.org/action_groups/sag.html.

✔ **RFID pure plays:** Vendors like ConnecTerra, GlobeRanger, and OATSystems emerged out of the early pilots sponsored by the Auto-ID Center. These vendors, along with some RFID hardware and software veterans like RF Code and Savi Technologies, offer products that integrate with RFID readers, filter and aggregate data, and even incorporate business rules. Some vendors have come out of device management for the Department of Defense, like Cougaar Software. These vendors are still in their early stages, but their involvement in pilots and RFID standards development has turned them into valuable resources for practical RFID middleware know-how.

✔ **Application vendors:** Driven by RFID mandates, application vendors like Provia Software, Manhattan Associates, RedPrairie, and SAP now offer software ranging from RFID-enabled applications for warehouse and asset management to more robust RFID middleware solutions with reader coordination, data filtering, and business logic capabilities. Many application vendors like Provia Software and RedPrairie have tackled this market by striking up relationships with specific middleware vendors. Other vendors, like Manhattan Associates and SAP, have built their own solutions.

✔ **Platform giants:** Vendors like Sun Microsystems, IBM, Oracle, and Microsoft are extending their application development and middleware technology stacks to handle RFID requirements. Each of these vendors is working to amass RFID experience and bring a strategic RFID middleware architecture — which leverages its standard application-development, data-management, and process-integration products — to market. These vendors bring unparalleled experience with highly scalable application platforms to the table.

However, the platform giants are going about this in different ways. IBM, for example, partnered with a start-up until it understood enough about the market and is now producing its own piece of software that directly competes with its former partner. Sun Microsystems took RFID as a strategic initiative and began developing the software piece in 2002. Microsoft has partnered with many companies in 2004 and opened up an application lab to showcase those solutions built on the .NET platform.

✔ **Integration specialists:** Similar to the platform giants, integration specialists like webMethods, TIBCO Software, and Ascential Software are adding RFID-specific features like reader coordination and edge-tier filtering to their existing integration broker technologies. These vendors offer extensive experience with high-volume data and process-integration scenarios and have an opportunity to capitalize on RFID adopters that have already invested heavily in their integration broker technology.

It's highly unlikely that any one vendor can solve all your RFID middleware needs. Each type of vendor offers one or more pieces to the middleware puzzle. The key is to assemble those pieces into a well-orchestrated solution that fits into your existing architecture. In the next section, I help you do just that.

Piecing Together a Middleware Architecture

To develop a network architecture for your middleware, you need to follow three key steps:

1. **Understand the RFID middleware architecture tiers.**

2. **Evaluate your existing middleware investments.**

3. **Prioritize your middleware needs according to your deployment plans.**

In the sections that follow, I walk you through each step in detail, with tips for customizing your architecture for your business and finding the middleware and middleware vendors that meet your needs.

No more tiers: Grasping the many levels of a middleware architecture

One of the key reasons for the hubbub around RFID is that it gives you visibility into what's going on across all points in your supply chain — from the receiving dock, to the production line, to transportation vehicles, and even to the retail store shelves. That's a lot of data coming from a lot of sources.

To process RFID data efficiently, middleware functionality can't be confined to a centralized data center somewhere in Nebraska. If it were, your network would come to its knees very quickly. Instead, middleware needs to be distributed with the right level of logic placed at the right location, or tier, in the architecture. I find it easiest to think about an optimal RFID middleware architecture in three tiers, as shown in Table 11-1.

Table 11-1	Three Tiers of RFID Middleware	
Tier	*What It Does*	*Where It's Located*
The Edge Tier: The primary function of the edge tier is to serve as the first line of defense from an overburdened network.	Very basic filtering to filter noise and superfluous data from the network, such as duplicate reads, which often still exist despite advances in reader technology. May also aggregate multiple reads into "packages" of data that can be passed up to local applications, rather than sending individual read information.	Close to — or even on — the readers themselves. In the past, this logic resided on separate boxes, placed as close to the readers as possible. As readers become more intelligent, they host this middleware logic themselves, nipping unwanted reads in the bud right at the source.

(continued)

Table 11-1 *(continued)*

Tier	What It Does	Where It's Located
The Operational Tier: The role of this tier is to do more context-based filtering that requires knowledge of other reads coming through the system.	Decides where to route the data — either to a local warehouse management system or up to the enterprise, for example. Raises flags when exceptions occur (like when a pallet tries to leave a distribution center without enough cases on it) using business-event management logic.Stores some RFID data in a database so that a monitoring application can track all traffic flowing through that site.	At individual sites, like warehouses, distribution centers, or retail stores.
The Enterprise Tier: The highest level in the architecture is similar to existing enterprise integration tools from vendors like TIBCO, webMethods, and so on. The goal of this tier is to accept data from the operational tier and incorporate it into enterprise-wide processes and/or applications.	Connects with common enterprise applications and data stores, like SAP or a centralized product information database. (Advanced systems will actually have process-management capabilities and some prepackaged logic for this task.) Communicates data to external business partners, like an advanced shipping notice that needs to be sent from a manufacturer to a retailer.	Oftentimes at one central data source where the information can be mined and acted upon for business decisions.

View these tiers as nothing more than a good guideline to help you think through the potential architectures that could work in your environment. Figure 11-1 illustrates one possible three-tier architecture, but every company's number-specific business requirements are different and shape the definition of the optimal RFID architecture. Again, the key is to build a flexible architecture that can support the right level of logic at the right location.

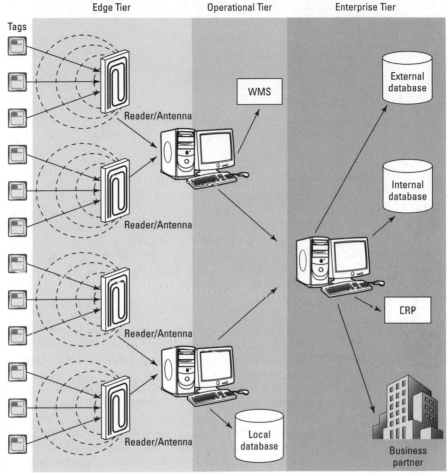

Figure 11-1: A multi-tiered middleware architecture will optimize processing.

Source: Forrester Research, Inc.

Taking stock of existing investments and skills

Now that you understand the different tiers of a complete RFID middleware architecture, it's time to figure out how you're going to cobble them all together. To avoid overbuying, the first step is to identify which tiers, if any, can be accomplished with technology that you already have inhouse. Chances are, you've already made some investments in other types of middleware, including

- ✓ **Application servers:** Examples include IBM's WebSphere Application Server, BEA's WebLogic Application Server, and Oracle's Application Server.

- ✓ **Integration brokers:** Examples include TIBCO's BusinessWorks, SeeBeyond's eGate Integrator, IBM's WebSphere Business Integrator, Microsoft's BizTalk Server, SAP's Exchange Integrator, and webMethod's Integration Platform.

- ✓ **Data management and integration tools:** Examples include Oracle's database, Microsoft's SQL Server, IBM's DB2 and/or WebSphere Product Center, and data-integration tools from Ascential Software.

All these technologies can play a part in your RFID middleware architecture. For example, existing integration servers — assuming they offer content-based routing, data transformation, application integration, process management, and good reliability and scalability features — can take on the middleware responsibilities that are needed at the enterprise tier. These decisions should be made after you have a good handle on what the infrastructure architecture will look like so that you can define your interrogation zones, reader devices (hand-helds, fixed portals, and so on) and communication and control needs.

Identifying what existing technologies you can use in your RFID middleware goes hand-in-hand with identifying how you need to expand upon those existing technologies. Here are some tips for ensuring that the new pieces of your architecture fit well with existing pieces and inhouse expertise:

- ✓ **Research which RFID vendors your existing middleware providers have partnerships with.** These partnerships are clues to which solutions will work well with your existing investments. Both TIBCO and webMethods, for example, have partnerships with OATSystems, which could provide the features needed at the edge and operational tiers. Most of the other integration broker vendors have similar partnerships.

- ✓ **Take stock of which technology skills your staff already has.** Knowing whether you are primarily a Java shop or a .NET (Microsoft) shop will help you narrow down your choices of vendors because each vendor tends to excel in one environment over the other. And, because most RFID middleware tools are still quite immature, your staff is bound to have to dive into the code and do some customizations.

- ✓ **Demand real-world use cases of your chosen hardware solutions with the various middleware players you are considering.** In a rush to grab market share, many middleware providers claim interoperability with *all* hardware, readers, printers, and hand-helds. I've deployed most of the major middleware vendors and found that every one of them has a chink in the armor somewhere. You usually don't find out about it until you try to do the install. This is also why it is critically important to install a test version in your or your partner's lab.

If you're not ecstatic about the investments you've already made in enterprise integration software from an EAI (enterprise application integration) vendor or have been considering a switch for some time, now's the time to make some hard decisions. EAI tools from vendors like TIBCO, SeeBeyond, and IBM can be a perfect fit to serve as the foundation for your enterprise tier. An interesting way of looking at this is with some of the newer Web-based integration companies like Grand Central as well. But make sure that you're happy with the EAI solution before you make it a cornerstone of your multitiered RFID middleware architecture.

Early bird or late bloomer? Prioritizing your middleware needs

RFID mandates — which each have unique requirements — have ignited immediate demand for a wide spectrum of RFID middleware functionality and have put the RFID middleware industry on the fast track. Still, standards bodies like EPCglobal are constantly updating middleware standards, and many of the vendors are working overtime and have adopted accelerated product release schedules to stay current with changing standards and get enterprise-class products to market as quickly as possible.

The bottom line: The market — and the products on the market — are constantly and rapidly evolving. This makes it absolutely essential to coordinate your RFID middleware investments with your expected RFID rollout timeframe. Understand the physics first and then design the middleware solution around your infrastructure and rollout for best results. The middleware space is only getting more and more competitive, so take your time and make the right decision.

Every company should lay out a long-term multitiered RFID middleware architecture. But the path to reach that end state will vary, depending on whether you're an early adopter who must get started immediately or a company that has more conservative RFID adoption timeframes:

✔ **Early RFID adopters** — particularly those that need to meet fast-approaching deadlines from retailers like Wal-Mart, Target, and Albertsons — should start with middleware products that have sophisticated reader coordination and data management tools as well as application features like EPC commissioning and track-and-trace tools. Also look for middleware vendors that have a strong commitment to standards because such a commitment not only ensures that your initial implementations hit the ground running but also offer the opportunity to integrate with a more scalable platform as RFID deployments grow.

After immediate mandates or pressures to get a specific application of RFID up and running are behind you, RFID adopters with large-scale RFID deployment plans should look to larger vendors that can provide scalable infrastructure and integration features. These vendors will provide the backbone for creating for a flexible, multitiered architecture that supports high-volume data and process integration scenarios.

✔ **Those with more flexible timeframes:** Consider your long-term application strategy. You have the luxury to plan several years out and create a middleware strategy that is as scalable as your enterprise architecture strategy. You also have the time to evaluate large-scale offerings from some new players in the RFID space who are highly experienced in the enterprise space — players such as SAP, Oracle, and PeopleSoft.

Getting the Most from Your RFID Middleware

With a complete RFID middleware architecture in place, you reap several immediate and on-going benefits that result from all of middleware's assorted functionality, including the following:

✔ Minimized network traffic through intelligent filtering

✔ Lower reader-management costs through centrally coordinated readers

✔ Immediate visibility to pertinent RFID data through routing, filtering, and track-and-trace tools

✔ Minimized on-going integration costs through standard APIs and prepackaged application integration tools

Well-architected RFID middleware can also enable more strategic opportunities that go way beyond these initial, rather obvious benefits — but only if you know how to unlock those opportunities. I've alluded to these opportunities a few times earlier in the chapter, but now it's time to come right out and say it.

The conventional goal of RFID middleware is to intelligently pass RFID data from readers to enterprise applications, which is a valid goal. But an arguably more exciting application of RFID middleware is to serve as the foundation for building new applications that take advantage of real-time, item-level data

CASE STUDY

Lessons from a real-life RFID middleware evaluation: Pick a deployment scenario

To assess the state of the RFID middleware market and see how the vendors stack up against each other, Forrester evaluated leading RFID middleware vendors using approximately 75 criteria that spanned everything from edge-tier features to enterprise-level process management. Forrester included 13 RFID middleware vendors in the assessment: ConnecTerra, GlobeRanger, IBM, Manhattan Associates, Microsoft, OATSystems, Oracle, RF Code, SAP, Savi Technology, Sun Microsystems, TIBCO Software, and webMethods.

The results were so varied across vendors that it was impossible to do one apples-to-apples comparison. Instead, Forrester actually separated the evaluation into two components: one that prioritized features essential for early adopters and one that focused on features most important for supporting longer-term, high-scale deployments.

You can purchase this document from Forrester at www.forrester.com/Research/Document/Excerpt/0,7211,34390,00.html.

and support new ways of doing business. The benefit of RFID comes from changing existing business processes. To do this, you must leverage the ability of RFID to create rules and actions in a true machine-to-machine communication format. The system should be able to do everything from inventory forecast cycles to creating pricing strategies.

Admittedly, RFID middleware alone can't provide everything that's needed to build this new breed of applications. For example, having a monitoring system that is separate from the data function is a critical component to keeping the RFID reader network healthy, as I explain in Chapter 14. But middleware can and will accelerate the innovation process.

Part IV
Raising the Beams for Your Network

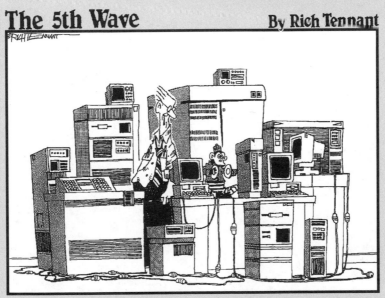

The 5th Wave   By Rich Tennant

"NOW JUST WHEN THE HECK DID I INTEGRATE THAT INTO THE SYSTEM?"

In this part . . .

*P*art IV is dedicated to long-term planning and adoption of the RFID network. I help you plan for and deploy your pilot and plan for its on-going impact on your organization. Going through some of the exercises and analyses in this part will help you decide on the scope and timing of your RFID deployment.

You also find out how to set up your own laboratory so that you can stay abreast of the latest technology, figure out how specific technology will fit into your environment, calculate the true cost of that technology, and generally avoid your stiff suit and tie because you've got work to do in the lab.

Chapter 12

From Pilot to Admiral: Deploying RFID Successfully

I've never been accused of being the best planner in the world, but I do know that having a solid project plan and a capable project manager is invaluable to the success of any critical technology project. Working through the plan is where order and execution meet, and in the case of a successful project, these are brought together by a person whose first career might have been as drill sergeant at Parris Island. If you've been around your company or the technology world for a while, you may have taken part in other major new IT initiatives or deployments. And now you're participating in planning and executing an RFID deployment. Congratulations!

In Chapter 1, I introduce the *Four Ps,* which outline the critical overall steps for a successful RFID deployment. Remember that you need to accomplish the first two Ps — planning and physics — before you begin a pilot deployment and then roll out to production. That's because RFID fundamentally changes the way that you count and capture inventory data. And beyond the warehouse environment, this kind of fundamental technology change ripples throughout your organization. The planning stage includes assessing your business touch points, which I explain in detail in Chapter 3, and understanding the physics of radio frequency and RFID hardware, covered in Parts II and III.

Your first foray into RFID is an exciting event — the greatest learning experience you will undergo. With this in mind, this chapter guides you through the pilot and into production. Here, I explain how you put your planning and physics knowledge to use in order to get the most out of your pilot, from creating specific milestones to conducting milestone debrief sessions. Looking back at some of the metrics, the design hypothesis, and the results helps you build an iterative process for redesign until the system is fully operational and ready for production. Dress rehearsals provide the opportunity to experiment, analyze, and correct any unforeseen problems before full RFID integration and improve the probability of a successful full deployment.

Creating a Pilot Project Plan

This section provides you with a template and format for planning your RFID pilot deployment project. By working from this template and format, you benefit from established project management methods and get your RFID deployment off on the right foot.

A project plan that takes into account all the areas of business affected by the RFID pilot deployment helps you track the project and keep it moving forward smoothly. Your project plan can be as simple as an Excel spreadsheet that has timelines and specific categories for the various areas impacted — like facilities, IT, and business processes. If you are lucky enough to have a copy of Microsoft Project (and if you know how to use it), your project plan can be much more sophisticated than a simple spreadsheet. You can find details about how to use Microsoft Project in *Microsoft Project 2003 For Dummies* by Nancy Stevenson (Wiley).

One handy project communication tool I always push for is a project Web site. Use the Web site to post the original project plan — including the risk log — as well as any updates so that all team members and stakeholders can access the information.

Any pilot project plan you set up must provide you with the mechanisms for outlining the requirements and tasks for your RFID pilot deployment. In addition, the project plan keeps track of important milestones, lists project participants and contributors, and serves as a communication tool used by all members of the pilot project team. So, if you follow my advice in the upcoming sections and don't skimp on setting up the right pilot project plan, you will develop an RFID pilot that will be successful and highly educational for your team.

If your RFID pilot deployment is like most pilot projects, a number of players have influence over, input to, or accountability for the project. This fact means that you'll be sharing the project plan among many people. And each person looking at that project plan is likely to make additions, deletions, or comments. Because your plan has this community aspect, it is critical to protect the plan integrity by setting up a *version control system* (a mechanism to denote the latest updated version) early on. The easiest version control system I've found uses the file naming convention *PROJECTNAME_DATE_VERSION*. So you might have a project plan filename that looks like SHANNON_02172003_1. Each person who makes changes to the project plan updates the version number and the date, and if there are multiple changes on the same date, the version number goes up accordingly.

Start with your major tasks and timeline

No matter what project tools you decide to use (from a simple spreadsheet to complex project management software), make sure that you include the following critical categories as primary headings in the plan:

- ✔ **Procurement:** This category indicates when you need to procure various pieces of equipment, like readers and cabling, and what lead time is associated with getting those items. So if you are custom making racks to hold the readers and antennas, working back from ordering the raw materials is critical.

- ✔ **Facilities:** You need to factor in changes to the facility that need to take place before you start installing the readers or cabling data communication. Such changes may include adding power at the various interrogation zones (where the RFID readers will be).

- ✔ **Business:** Identify what processes have to change after the system is deployed. For example, you might want to specify that the receiving or picking process changes and that training takes place so that employees understand those changes. If you plan on using hand-held readers, you may need to allocate a certain number of days for training on these items. If your current processes involve rough handling of items, you may need to ensure that those processes change so that RFID tags that will be applied to those items aren't damaged.

- ✔ **Physical:** Include all the things in the physical environment that need to change — from forklift routes to conveyor sizes — in order to take advantage of RFID.

- ✔ **Systems:** Critical to the success of RFID is having systems that can take action on all this real-time serialized data. If you completed the necessary planning, you know how to use the data as the RFID network creates it and what applications it will integrate with. Your project management tools need to break down the tasks involved in producing the desired end result.

You can construct the project plan on a single worksheet with the five headings just described. Figure 12-1 and Figure 12-2 show examples of an Excel spreadsheet work plan that tracks a pilot project. At the bottom of

Figure 12-2, you can also see a section devoted to key milestones and where they fall within the project. Important key milestones for an RFID pilot deployment may include the following:

- ✔ Selecting the RFID team and assigning roles and responsibilities
- ✔ Understanding the details of any mandates or partner requirements
- ✔ Obtaining current metrics for measuring supply chain process and efficiency
- ✔ Becoming educated on RFID in general

Deliverable tracker

After you lay out the project tasks and timeline and document the critical milestones of the plan, you can turn your attention to the dependencies and deliverables associated with each of the sections. Outline the deliverables and their due dates and then add this list to your project plan as a *deliverable tracker*. In project-management lingo, the deliverable tracker is an explanation of resources that are required to successfully complete the project.

A deliverable in an RFID pilot deployment can be anything from installing cables at specific locations to training personnel on equipment operation. Include the deliverable tracker in the same spreadsheet or Microsoft Project file as your plan's tasks and timeline.

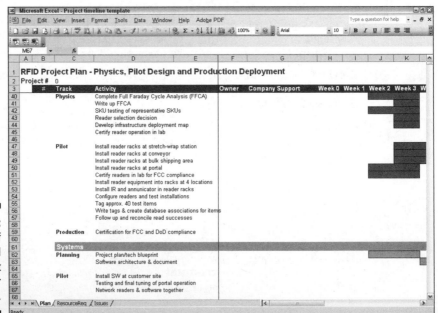

Figure 12-1:
The top half of an Excel spreadsheet to track your RFID pilot.

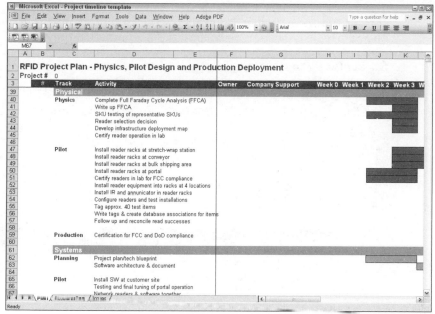

Figure 12-2:
The bottom
half of
the project
work plan
in Excel.

You can use the items on your deliverable tracker as checkpoints for your pilot deployment. I recommend reconciling the deliverable tracker with the overall project plan (timeline and milestones) on a weekly basis.

There's always an issue with you: Tracking and resolving problems

A third component of your project plan is a mechanism to track any issues (that is, problems) that may come up as your project progresses. Because our friend Murphy is always rearing his head to enforce his law (and why don't Sully and Malone have a law like Murph does?), having an *issue tracker* in place within your project plan is essential. A good issue tracker helps you quickly resolve issues, get to the right people to take action, and know whether what you're doing is the right solution. The issues tracker explains the following:

- ✔ **Who brought up the problem,** in case resolving the problem calls for further explanation or clarification.

- ✔ **A definitive timeline to resolve the issue,** which is essential for keeping the project moving forward (a prime reason for having a project plan).

- ✔ **A single person to take accountability for resolving the issue** because, after all, you want the issue resolved, and making someone responsible for the resolution is key.

In the project document, include a page that you can use to track issues, such as the one shown in Figure 12-3. To make sure that items are being acted on and people are getting the support and information they need, the project manager checks the issue tracker weekly. If problems linger in the issue tracker for a number of weeks, the project manager knows that something probably either requires higher-level attention or has the potential to jeopardize success in the long run.

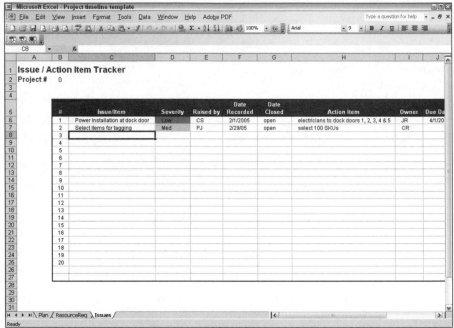

Figure 12-3:
The issue tracker for any unforeseen problems.

There is no 1 in team (but there is an M and an E)

A team roster is the final part of the project document that is critical to keeping your RFID pilot deployment on track and well organized. This roster, as the last sheet of the project plan, includes critical contact info for all involved parties. Make sure your team roster includes names, departments or business affiliations, e-mail addresses and telephone numbers or other such contact information, and a description of each team member's role in the project. Having a complete roster ensures that the project manager can get ahold of the right people at the right time.

Factors for a Successful Pilot Test

Having a beautifully constructed pilot project plan is a real plus, but other not-so-easy-to-document factors can give you an extra boost toward accomplishing your pilot project goals. In fact, seven critical success factors can help you deliver your RFID pilot deployment on time and within budget. These seven factors (described in the upcoming subsections) are

- Clearly defined scope
- Experienced project manager
- Key executive support
- User involvement
- Specific measurements and metrics
- Risk mitigation
- Phased approach

Clearly defined scope

Projects tend to fail due to poorly defined requirements and scope creep. (No, scope creep isn't a guy who skulks around clutching a mouthwash bottle.) *Scope creep* happens when the original, well-defined project gets more and more requirements layered onto it — add another read station, send data to more applications, and so on. To avoid leaving room for ambiguity in your RFID pilot deployment plan, document the following aspects of your pilot project:

- **Scope:** How much, for how long, and so on
- **Requirements:** What data to collect, what equipment to test, and so on
- **Goals:** For example, what percent of reads you require to be better than the current system or specific results you are after, like eliminating two jobs per shift.

Make sure that your project's goals and timelines are realistic!

Have all these specifications well-honed before you turn the project over to an experienced project manager and begin soliciting executive support. You do much of the work to define the scope when you prepare your business case for RFID deployment; see Chapter 15 for details.

The specification document you create is your road map to determining what needs to be done and how you measure success. As you detail your requirements, think beyond the pilot deployment and consider *scalability* (that is, how to adjust the system's workable size) for the RFID system that will ultimately roll out to different parts of your business. Always design with the end in mind.

Experienced project manager

The project manager is largely accountable for the success of your RFID pilot deployment. Make sure that your company chooses a project manager with the necessary skills and experience in project management methodology and planning. The project manager's ability to keep track of the plan and all the players is crucial to long-term success.

After you recruit a kick-butt project manager and clearly define your project scope, you have a suitable package to present to key executives for support of the project.

Key executive support

In Chapter 15, I show you why it's so important to assemble the right mix of people to develop a winning approach. Be sure to assemble a cross-functional project team with a shared vision for RFID deployment and foster cross-area executive sponsorship and support. Present the project plan, timeline, and metrics to key executives for sign-off and agreement before putting your plan into motion.

User involvement

The entire corporation or agency must be engaged in the vision and execution of the pilot test or full RFID deployment. Obtaining input from key stakeholders facilitates buy-in and helps you refine your pilot project's scope, requirements, and measures of success. Key stakeholders from your business commonly include the manufacturing team, logistics, finance, and IT. Chapter 15 talks about building a team of these key players, and Chapter 16 offers more tips for achieving company-wide support for an RFID project.

In some cases, you might also need to engage your suppliers and customers because they, too, can help support the vision and execution of the RFID pilot project by deploying their own RFID networks and either sending data back to you in real time or shipping you products with RFID tags already on them.

Specific measurements and metrics

Defining the criteria for a successful RFID pilot deployment and telling how you intend to measure results are imperative. Criteria for measuring the success of your inventory management system can be as simple as collecting data on the number of cases lost each month or as sophisticated as tracking the average timeline through your entire distribution channel.

For your RFID pilot, document your "as-is" processes and establish specific performance benchmarks under those existing processes. Then, during the pilot deployment, measure the performance of your "to-be" processes and compare against the benchmarks to determine whether you met your success criteria.

Look to Chapter 14 for more information on establishing benchmarks and defining system metrics.

Risk mitigation

During the initial project-planning phase, produce and maintain a risk log and an associated action plan to mitigate each risk identified. For example, a potential risk for an RFID pilot project is not being able to do anything with RFID or EPC (electronic product code) data when your RFID network begins producing that data in real time. An associated action plan might be to plan which business applications receive data, create rules for what to do with the data, and then measure the effectiveness against your existing systems.

 Many people, when deploying an RFID pilot, set up an entirely separate stand-alone system and compare its results with the results of the existing system. This process is especially useful in security or asset-tracking scenarios in which manual recording of property is fraught with human errors. A fully automated RFID system compares quite favorably with the old-fashioned manual systems.

Publish your risk/mitigating action log for use by your team and your stakeholders. Knowing what action you plan to take should the worst happen saves time and helps the pilot project team handle potential showstoppers.

Phased approach

You want to design your RFID pilot deployment test to minimize the impact on your production environment while simulating your real operation. Using a phased pilot approach — that is, a series of tests that grow in scale — can help you achieve this goal.

Before you implement RFID for any asset tracking, shrink reduction, or security pilot, be sure that you have accurate numbers for the current state of affairs. Having an accurate baseline metric gives you something to compare the RFID results to. In security applications, having your current system running alongside your pilot RFID system can't provide an accurate baseline because just knowing another system is monitoring behavior can make people act differently and improve statistics.

Your series of tests, or pilot phases, might look something like this:

1. **Set up an RFID interrogation zone at only one location. For example:**

   • **A dock door zone that duplicates the work of your current bar code scanning system:** Let the RFID system run for one month and reconcile your results with those of the bar code system.

   • **A shrink-wrap station zone that gauges accuracy of case-level reads on a full pallet of product:** Use the average read percentage as data for your ROI analysis.

   • **An asset-tracking zone in which only one type of object is tagged, such as laptop computers:** Compare RFID data of that class of objects (laptops) with sign-in and sign-out sheets, particularly for secure areas, and reconcile the results.

   I explain the general steps for setting up interrogation zones and training users in Chapter 13.

2. **Using your understanding of RFID physics, redesign that one interrogation zone until it's right.**

   In your team meetings at the pilot stage, focus on the following areas:

   • **To make sure the technology works properly, compare various components.** How does a circularly polarized antenna compare to a linearly polarized antenna? What happens if you change the antenna angle by 45 degrees? What if you move the transmitting antenna back a bit and leave the receiving one where it is? The pilot helps you develop standard configurations that become important in your production environment. A team member should document all these technical attempts and changes so you can look back as you do future deployments and know what expected results of various changes should be.

   • **At the debrief sessions, you may have to address what business processes need to be done differently and test those changes in the pilot environment.** For instance, say that one of your goals is to get 100 percent case reads as pallets pass through a dock door. If the technology is correctly set up and optimized and you're still not getting 100 percent, you can try changing the business process. Making the forklift driver stop for 10 or 15 seconds may give the system enough time to meet your goals.

At the pilot stage, if you decide to change some of your existing business processes, it is critical to design the system so that those changes can be incorporated and then communicated back to the humans interoperating with the RFID network. This could be as simple as setting up a red light/green light timer for the dock door or a buzzer after the 10 or 15 seconds is up.

3. **After you are successful with one dock door portal, conveyor, or shrink-wrap station, start your next deployment or phase of the pilot next to the first one.**

   Putting the two pilot deployments side by side enables you to understand the challenges inherent in a multiple reader environment and helps your team learn the nuances and importance of proper configuration and setup of the readers.

By phasing the pilot, you gain more focus and control, using your measures of success for *gating* (making the go-ahead decision to move on to the production phase). In your final pilot phase, you move on to a live trial test — that is, a production trial with live product and data.

Moving from Pilot to Production

Moving from a pilot into a production environment is the difference between grade school and college, involving a lot more complexity, not as much control, a whole lot of unknowns, and the occasional embarrassing outbreak of acne. After your successful completion of the pilot and initial deployment, the project debrief is a critical step toward moving from pilot to full deployment.

Getting the most of your pilot data: The project debrief

For a project debrief, you need to follow these steps to cull the important lessons from your pilot data:

1. **Collect as much relevant data from the test as possible.**

   The most important data is, of course, how the RFID methodology compares with what was being used before (bar code, manual systems, and the like), but also includes things like setup time, equipment lead time, costs associated with the project, and so on.

2. **Organize the data in the categories you originally set out for the project plan.**

   For example, your IT category might show the impact of increased storage needs, or another system to be added to a view in a network operations center.

3. **Schedule the team together for a full-day working session. During this session, the team evaluates**

   • **How well the original goals of the RFID project were met**

   • **What the next steps in the RFID process are:** A valuable way to do this is to incorporate the actual learned data into an ROI analysis. Based on this analysis, rate the areas impacted the most from the process, reassess potential benefits, and refine the next steps.

4. **Present all the distilled and summarized information, including next steps and their budgetary requirements, to senior management.**

   Trust me: CEOs like to have as much work done for them as possible. Give them the good, the bad, the ugly, and what's next for the project and why.

After you understand how your enterprise is going to react to an RFID system, you have a better understanding of what issues you should focus your deployment on.

Tips for a successful production system

The goal for your production environment is to make the transition to RFID as effective and low-maintenance as possible. The best way to do this is to

✔ **Standardize your tag type.**

✔ **Limit the number of different readers you use to one or two.** On rare occasions, you may need three different kinds of readers to achieve optimal results, but usually two different types can cover most applications.

If you design with the end in mind, you set up the readers to perform well in a system environment during your pilot. So at this stage, during your full deployment, you are able to take two dock door configurations and set them up next to each other with no interference issues, no ghost reads, and great accuracy.

Chapter 13

Getting Set to Administer and Maintain Your System

*R*FID makes all transactions automatic, doesn't it? In an ideal world, yes. But then, in an ideal world, all the work is completed while you're at home having breakfast or bungee jumping out of a hot-air balloon.

By the time you get to your pilot or production deployment, you may be tempted to think that you're almost done, that the machines will be taking over soon. Well, RFID enables more efficient operations, but we humans must still know what those operations are and how to execute them. You *have* come a long way, but setting up a pilot or production deployment is just another step in the learning and implementation process that you have undertaken.

Your pilot deployment is your opportunity to put the pieces together and see how your efforts in the lab (see Chapter 8) pan out in your actual manufacturing environment. As I stress throughout this book, an understanding of physics is the cornerstone of a solid foundation in RFID (see Part II). In this chapter, you put that understanding to use in order to set up the equipment correctly. A pilot is an iterative process: Based on what you discover during your first pilot, you can continually redesign the system until you're ready for production rollout.

Chapter 12 explains how you move from pilot to production from a project management perspective. This chapter explains the hands-on information you need to set up equipment and processes so that you capture the necessary data accurately and reliably. Your staff also needs to understand how to use the RFID systems so that you achieve maximum benefit from RFID, so I offer tips for training staff in this chapter, as well.

Configuring and Setting Up Tag Readers

Someday, readers will be completely self-configuring: They'll automatically know where to read tags, configure themselves for optimum performance, and download all the right applications to properly handle the read data. Until that day of RFID nirvana comes, however, you need an expert to locate and configure readers manually by using sound principles of physics and a deep understanding of the various types of hardware available.

If you're in the pilot stages, focus on finding the right hardware and playing with various configuration classes — one configuration for the dock door, one for the conveyor, and so on. If you're ready for full production deployment, you know from your pilot data what configuration classes to use for various locations and what hardware is optimal for the entire planned RFID network. You just need to carry out the implementation accurately.

Your staff relies on the readers being properly configured so that they (your staff) can just do their jobs. So it is imperative that the readers — whether fixed-location, mobile, or hand-held — are properly configured and ready for use when the shift begins.

Before you begin

As I say throughout this book, you need to design with the end in mind. That's why proper reader configuration begins with the planning stages. Before you begin setting up the actual interrogation zones, you and your RFID team need to figure out the following:

- **Where you want to set up interrogation zones to read tags** (Chapter 3)

- **Where and how power and network connections can be made available to the readers:** In addition to actually installing cables, you need to know how to protect cables connecting a reader to its antennas. You need a plan to ensure that no one line of the power feed is overloaded. Similarly, the network must not be overburdened with traffic, and the use of Ethernet for network communication limits the maximum distance between repeaters or routers. (Chapter 3)

- **Which tags you want to use** (Chapter 5)

- **The right reader for the required job** (Chapter 10)

- **The total area the readers need to cover:** Basically, this means that you need to know how many antennas you need to get the desired performance, without interfering with other systems. (Chapter 10)

Stepping through a reader setup

Although each interrogation zone setup is unique, this section gives you a general idea of the steps for putting the puzzle together and offers tips that can help along the way. Here are the basic steps involved:

1. **Mount the reader (see Chapter 10 for details).**

2. **Attach the reader antenna in the appropriate location and make sure its orientation is correct.**

   When locating new antennas, take care to minimize potential RF interference. Make sure directional antennas aren't pointed directly at one another, regardless of distance, unless their operation is synchronized (that is, if you have configured the reader to cycle through the antennas in a specific order or you are using software that coordinates the cycling of antennas). Unsynchronized operation can lead to unpredictable behavior due to tag confusion and reader interference. See Chapter 10 for more details about antenna setup.

   If this is your first pilot setup, remember to check for differences between your lab environment and your actual environment. Laboratory environments are, by definition, experimental. These differences might impact installation of the reader and antennas and might impact the RF performance of the system. You may need to reevaluate the system in its actual installation, which of course is what the pilot is all about.

3. **Mark the correct location and orientation of the antenna by drawing permanent lines on all the brackets (if possible).**

   By marking the correct setup, you can quickly identify and properly orient an antenna, which is critical to the maintenance of the system. Plywood templates may also be used to simplify the reorientation of the antennas.

4. **Run all the needed power and cables to the reader and power up the reader once the antennas are connected.**

5. **Test the interrogation zone for path loss (see Chapter 10).**

6. **Tweak the reader's configuration settings as needed.**

   Work from the settings you determined in the lab (see Part III). Although you may be tempted to tweak the configuration as needed through trial and error, it's more effective to work from the laws of physics and the known behavior of the frequency and system you are deploying.

   Look out for configurations that leave overlapping interrogation patterns from unsynchronized antennas. Such overlaps can cause unpredictable system performance due to tag confusion.

I discuss the details of configuration files and how to store them in "Creating configuration classes," later in this chapter.

7. **Set up an uninterruptible power supply (UPS) for each reader.**

   A UPS protects your data in the event of a power outage.

8. **Your business process may require that, in order to take full advantage of the capabilities of RFID systems, you set up a "verification tunnel" to ensure that all tags on a pallet load or contained within a reusable container are read.**

   To set up this tunnel, you need to find out what your system level performance and hardware requirement is. For instance, Wal-Mart demands read performance on a 10-foot wide dock door with varying antenna configurations. Today those would be deployed over a Matrics AR-400 reader (which is now being produced by Symbol Technologies). If you want to verify that Wal-Mart can read your tagged items, you can build a stand-alone 10-foot-wide portal and set up Wal-Mart's configuration. To verify performance, have your forklift driver go through this portal and receive a visual cue, such as a red light or green light, to determine whether he should load the pallet for shipment to Wal-Mart.

   Verified reads may be checked against the manifest to ensure 100-percent order delivery. This is a critical compliance step when the financial systems are integrated with the detailed delivery of products.

9. **If you use hand-held and mobile readers, don't forget to set up designated charge stations where users recharge the batteries on these devices, and also upload any necessary data to the central system if there is no wireless communication.**

   These charge stations must be both convenient and protected from physical damage.

10. **Set up and test monitoring systems for both the users and the system administrator.**

    The only way that you can have 24/7 support within an RFID network is to have the readers actively monitored and constantly checked for performance. I explain monitoring in Chapter 14.

Creating configuration classes

When I say "creating configuration classes," I'm not talking about teaching RFID to your kids. I'm talking about creating a standard setup for the various areas where you'll set up RFID interrogation zones. During your pilot, you determine how many configuration classes you need and what the right setup is for each one.

To the extent possible, try to create a limited number of configuration classes so that adding new interrogation zones to your RFID system is as simple as identifying its class, pushing the configuration out to the reader, and adding it to the monitoring system.

Configuration classes are stored in a database and look something like Table 13-1.

Table 13-1	Example of a Configuration Class
Field in the Database	*Value of the Field*
Configuration Class #	1
Configuration Name	Receiving Dock Door
Reader Manufacturer	Symbol (formerly Matrics)
Reader Model	AR-400
Firmware Version	2.1.00
Antenna 1	CP
Gain/Power 1	0.75
Antenna 2	CP
Gain/Power 2	0.75
Antenna 3	CP
Gain/Power 3	1.00
Antenna 4	CP
Gain/Power 4	1.00
Scan Period	Polling
Tag Filters	EPC Inclusive
Low Range Value	xx.xx.xx
High Range Value	xx.xx.xx
Protocol	Class 0
Network Attachment	TCP/IP
IP Address	192.168.2
User Access	Jetson, Rubble, Fudd
Password	RF4dummies

Here are some tips for creating configuration files:

✔ Store the reader configuration file in a secured database that has limited access.

✔ Develop the configuration file in a program that can interoperate with any open database connectivity (ODBC) database because many monitoring, management, and middleware programs may end up accessing the configuration files.

✔ Make sure you fully back up your configuration files to tape or disc once a month, and do incremental back-ups at least weekly, but preferably daily. Daily incremental back-ups are particularly important as you build out your network and as things change daily.

Getting the Digits

One of the biggest benefits of RFID over bar codes is the capability to uniquely number every item. Typically, you encode this number into each tag just before applying the tags to objects.

Most of the deployments I have set up so far encode tags with an RFID label printer. Zebra, Printronix, and Paxar bar code printers can have an RFID module embedded in them that allows them to work as both an RFID label printer and also as traditional bar code printer. The module broadcasts a special command with a discrete number (usually an EPC number) through the transmit (Tx) antenna, and that number is "written" onto the tag. Some of the deployments have used the readers to directly commission (write to) the tags. The operation is always controlled by some form of software — see Chapter 11 for more about the issues surrounding middleware features. Whether you use an RFID label printer or write to the tags directly through the readers, the technical aspect of sending a specific RF signal out the antenna with a write command is always the same.

Unique item identification has tremendous benefits, but assigning and managing unique item-level identification can be burdensome and difficult if not done properly. Having a numbering structure and a system for assigning numbers can simplify the allocation and management of unique numbers. The following sections explain proven methods for managing numbers.

A simple hierarchy for assigning numbers

All numbering schemes in use — whether for open RFID systems, such as the ones that enable retailers and suppliers to share information, or for closed systems, like the ones libraries use — rely on some sort of numbering hierarchy. All you need to do is understand the hierarchy so that you can use it to your advantage in assigning unique identifiers to all your items.

The general identifier of the electronic product code (EPC) is a good example: The general identifier EPC comprises three numbers whose combination uniquely identifies an item. Those three numbers are the manufacturer number, the product number, and the serial number. EPCglobal, the manager of the general identifier EPC, assigns a unique manufacturer number (also called an EPC Manager Number) to each of its member organizations. Each member organization is free to assign a product number (also called an Object Class) to each of that member's products. Similarly, each member is free to assign a serial number to each item manufactured by that member. Here's how you might take advantage of this hierarchy to uniquely identify an item:

- **Assign the same product number to all products with the same properties.** Thus, all 16-ounce blue bottles of Fabulous Fiber are assigned the same product number. All 24-ounce bottles are allocated a different product number. This is the basic approach used today by most product manufacturers.

- **Assign each item in a product line a unique serial number.** In this way, you can distinguish one 16-ounce blue bottle of Fabulous Fiber from another bottle just like it.

Serial numbers are new to the fast-moving consumer goods market, but they have been in wide use in the electronics, automotive, and aerospace industries, among others, for a very long time.

Allocating unique numbers across many lines and locations

Because all products of the same type receive the same product number, the serial numbers must be maintained in order to enable unique item identification. When a product is manufactured on exactly one production line at exactly one location, assigning unique serial numbers to every item manufactured is straightforward. A central numbering allocation authority may be used. Difficulties arise when multiple lines, potentially at multiple locations, are used to manufacture the same product.

When multiple manufacturing lines are used to manufacture the same product, a centralized numbering authority is more difficult to manage. The centralized authority must be consulted every time an item is produced, requiring that it be accessible over the network whenever production runs are in process. A centralized authority that is always reachable and is able to assign numbers at production speeds can efficiently use the numbering space. Additionally, careful details must also be maintained about which line an item was produced on.

An always-accessible central numbering authority isn't practical for many companies. These companies can use an intelligent hierarchy imposed on the serial number allocation to decentralize and make feasible the allocation of unique serial numbers across all manufacturing lines.

Here's a simple example of a decentralized, hierarchical approach to allocating serial numbers. A range of serial numbers for each product is allocated to each manufacturing facility. Within a facility, a range of numbers from those allocated to the facility is allocated to each line producing a particular product. In this way, the serial number is effectively subdivided into a facility number, line number, and subserial number, in which the allocation hierarchy is maintained between facility number and line number.

Remember to assign a line number for hand-applied tags that replace damaged or nonfunctioning tags. You need to look at your business processes to determine the best way to incorporate this.

Applying Tags to Objects

Eventually, all tags will be placed on products automatically, either on the packaging prior to receiving it or on-line within your facility. Until then, tags need to be manually placed on items.

When you apply tags to objects (or train staff to do so), you need to keep two important factors in mind:

- ✔ **Be sure you handle tags in a way that doesn't put stress on the tag's connections and parts.** Broken tags don't communicate well (if at all) and thus make your system less effective. See the sidebar, "Armor-plated tags," for details on emerging solutions to these vulnerabilities.

- ✔ **Place the tag in the optimal spot on the object.** During your testing (see Part III), you determine where the tag antenna can best couple (communicate) with the reader antenna. Some of the early adopters have already started printing tag outlines on their case boxes so that workers applying them know exactly where the tag goes and what the correct orientation is.

The following sections explain how you can work around these two factors as you apply the tags.

Applying tags without breaking them

Manually manipulating tags always has the potential hazard of physically damaging them. You can minimize the physical stress placed upon the tags by always following these steps when applying tags by hand:

1. **Place the label face down on a flat surface with the backing paper face up; the face of the label is therefore down on the table or flat surface.**

2. **Peel the backing paper away from the label while keeping the label flat on the surface.**

3. **Without bending or twisting the label, lift it from the flat surface.**

   Care must always be taken so as not to bend or twist the tags. Bending or twisting a paper tag places stress on the connections between the chip and the antenna. This stress may crack the connecting material, often a conducting glue, or it may break the connection completely.

4. **Position the label over the correct location on the object and smoothly apply it to the surface.**

   Similar care must be taken when pressing the label onto its final resting spot. Excessive pressure on the silicon chip can crack it, or, more likely, damage the connections between the chip and the antenna.

North by northwest as the corrugation travels: Orienting tags on objects

Your testing determines the best location and orientation to read a tag (see Chapter 9). When you set up your pilot or production system, applying the tag in just the right spot is particularly important so you know that the limiting factor is not the tag or the tag placement. Linear antenna tags, such as a simple and commonly used dipole antenna tag, which must be properly oriented on the package to ensure readability, are especially vulnerable to misreads caused by incorrect orientation.

Armor-plated tags

With all the talk about nanotechnology, you'd think someone could come up with chain mail for RFID tags to keep them better protected from cracking and breaking. Cracked connections cause two difficulties:

- The first difficulty is decreased performance for the tag. Although the tag might still operate, it will harvest energy less efficiently and communicate with the reader less efficiently. Often, this means that the tag is successfully commissioned or written to, but when it gets to your verification portal (or worse yet, your client's), it may not work at all.

- The second difficulty is that future shock or vibration may extend the crack, eventually completely breaking the connection.

Most of the tags in use today are vulnerable to cracked connections because they're made using a flip-chip technology. In the *flip-chip* process, the chip is "flipped" onto the antenna inlay with a special adhesive, which is less durable than a directly soldered connection. You can find companies that make soldered connections to antennas. There is even a company called Appleton Industries that adds a special high-durability coating to tags in an effort to make them more durable.

When the tags are applied manually, a simple diagram illustrating the proper tag location and orientation helps operators apply tags correctly. Figure 13-1 shows an example of how a box might be premarked for a production worker.

RFID tag here

(01) 1061414100734 6 (21) 2

Figure 13-1:
A preprinted box for manual application.

Sending Objects through Your Business

With the unique identifiers loaded into your tags and the tags applied correctly to the objects, you're ready to see how the whole system works together. During your pilot, you may need to assess and adjust your business processes so that tagged objects are oriented correctly as they move through your manufacturing line and to ensure that your equipment isn't damaged.

Lining up tags and readers

As you send items through manufacturing, remember that some items, particularly metal-wrapped products, allow an item's tag to be read from only a single direction. This is truer for some tags than for others. Physics once again rears its unwavering head.

You need to consider, in addition to the materials in your object, the physics of your RFID equipment. For orientation-sensitive items such as linearly polarized tags and readers, the orientation of the tag to the reader is crucial to reading the tag reliably. A linearly polarized reader antenna can read a linearly polarized tag antenna only if both antennas are oriented in the same plane. Practically, this means that if the reader antenna is oriented vertically, the tag antenna must be oriented vertically as well. Linear tags that are oriented horizontally will not be read by vertically oriented linear reader antennas. In fact, if you take a linearly polarized antenna, hold a linearly polarized tag in front of it, and rotate the tag 90 degrees, you can watch it disappear as far as the reader is concerned. (Don't worry; it will still be in your hand or on the box, but the reader won't see it. This is physics, not magic.) I explain the rules of physics that govern this behavior in Chapter 5.

Here are some tips for ensuring that objects are properly oriented when presented to the reader antennas:

- **If the boxes are stacked a particular way in the pallet,** during your tag testing, evaluate for a location that faces outward (usually the end of a box). If you know the minimum performance threshold or use the ODIN tag performance index (TPI) to determine the minimum threshold (for example, a TPI of 70 percent may be required to meet your mandate), you can try to test for a 70-percent result on the side of cases that face outward on each pallet. See Chapter 9 for more on the ODIN TPI.

- **On a conveyor belt,** side application is usually the easiest way to ensure correct read because the reader antennas are likely to be mounted along the sides of the conveyor. Mounting tags on the top works for conveyor alone but can be very difficult if you need to read cases on a stacked pallet later in the process.

Just like the neonatal ward: Handle with care

As I mention earlier in this chapter, damage to either a tag's chip or its connections to the antenna renders the tag inoperable. For labels that are located on the exterior of the packaging, the chip is vulnerable to impact from neighboring items.

If you run your finger over a label tag, you can feel the chip through the paper packaging as a very small bump, usually in the center of the tag. Being very small, the chip is surprisingly resilient to breaks; however, it is not indestructible. Because the chip protrudes beyond the label (hence the bump), it is vulnerable to

- **Impacts from several angles.** The impact from hitting a neighboring item may be sufficient to break the chip.
- **Being caught and twisted in conveyor belts and other equipment.**

During your pilot, identify whether these vulnerabilities are affecting tags by putting a series of pallets through an end-to-end test with your trading partner. Where the pallets are loaded, record the condition of the RFID tags on the cases (a digital camera up close works great for this) and then get to the partner's distribution center or store when your pallet arrives and record any areas that might have been impacted in transit. You should be able to clearly identify areas on pallets and cases that have been rubbed, bumped, or torn from a simple visual inspection.

If you want to get more involved than a simple visual inspection, some universities, such as Michigan State, have special equipment to simulate travel in the back of a truck — bumps, rattles, shakes, and all.

To correct the problem, figure out how you can change business processes — handling packages roughly should be at the top of the list of things to change. This is not to say that packages need to be treated with kid gloves. Certainly not. Just keep in mind that commercial from the 1980s in which gorillas throw luggage around an airport baggage handling system — a perfect example of the line being crossed while at the same time the rules of personal grooming being violated.

School's in Session — Training Your Staff

After you set up your system for a pilot or production deployment and make sure it is configured properly, you need to train your staff to use the equipment. Although they don't need to be RFID experts, the people who interact with the system on a daily basis must know how to do the following tasks properly so that your RFID deployment is successful:

- ✔ Start the system (if it isn't always on and operators need to start it manually)
- ✔ Use software applications (such as which buttons to push and when)
- ✔ Use the hardware (such as the point-and-shoot of hand-held readers)
- ✔ Apply tags to unlabeled items
- ✔ Orient tagged items as they move through your business so that the system achieves the best read performance

All operators must be trained to use the system. Even the simplest and most obvious of tasks, such as pushing a red button, may not be so simple or obvious while using the system.

In addition to knowing how to use the system, your staff must be supportive of the system. Teaching them the personal benefits of an RFID system is just as important as teaching them the benefits to the company.

Starting readers manually

Most fixed-location readers are either always attempting to read tags or begin reading automatically when some event occurs, such as a dock door opening.

However, when a reading location is seldom used or otherwise requires that the operator manually start the fixed-location readers, a simple, detailed process must be in place to ensure that the readers are operating before the operator expects tag reads to occur.

The best way to start a seldom-used reader is to have a terminal or workstation in close proximity to the reading station so that the operator simply presses the Start button and waits for the green "working" light.

Most installations don't have a co-located terminal and have a more complex start-up process. The operators must be trained in the appropriate process. In addition to the training, post the steps next to the reading location for those times when a novice operator is the only one available. Typically, the basic set of steps resembles the following process:

1. **Make sure antennas are connected to the RFID readers.**

2. **Make sure the equipment is plugged into its power source.**

3. **Turn on the equipment (a special sequence may be required).**

4. **Check that the antennas are correctly oriented and have not been moved.**

5. **Wait for the green "working" light.**

6. **Begin moving tagged objects through the read station.**

Although ensuring that the equipment is plugged in may seem obvious, it is surprisingly not when someone's actually in the field trying to debug a non-working system. You may have experienced this phenomenon in your own home: Say your lamp is usually plugged in, but someone unplugs it in favor of a drill, a fan, or some other device that for a short moment might be more important. Of course, people rarely plug an appliance back in when they're done, so whoever tries to turn on the lamp later that day ends up scavenging the house for a new light bulb before realizing the lamp was simply unplugged all along. In the warehouse, it's a little embarrassing to realize a reader is just unplugged while you're on the phone to your system administrator or technical support, so remember to include Step 2 when you post the steps on the wall.

Identifying and responding to missed reads

Using an RFID system includes knowing when RFID reads are to occur and, if the process requires it, verifying that the reads occurred. The operators must be trained to identify a proper read, and the system must provide this feedback to the operator. For fixed-location readers, colored light stacks, such as red, yellow, and green lights, provide usable feedback to the operators. Figure 13-2 shows just such a light stack in use. For hand-held readers, the ubiquitous "beep" has proven effective.

Red/green light

Figure 13-2:
A light stack that provides a green light signal if an item is successfully scanned.

Regardless of the feedback, the operator must be trained on the process to obtain the desired level of benefit from the system. For fixed-location readers, the appropriate actions based on system feedback (or the lack of feedback) may be to slow down, or even stop, the movement of tagged products until the desired feedback occurs. Here's how operators might respond when a reader doesn't produce the desired feedback:

- ✔ **If the system is designed to allow multiple reads of an item moving through a read zone,** the operator must be trained to repeat a process until reads are confirmed.

- ✔ **If the system cannot handle multiple reads** because, for example, a tag's location within or outside of a caged area is determined by the number of reads, the operator must be trained not to repeat the reading process. Another process, such as accessing a computer terminal, must be used to verify reads or to reset the system so that the read process can be repeated.

Reinforcing processes versus changing them

It's human nature for people to always follow the simplest set of procedures that achieves their ultimate goal. You'll likely discover that operators will circumvent or ignore official procedures that get in the way of these simpler operations.

Perhaps the best illustration of official procedures being circumvented comes from the use of bar codes on patients and medicines in the hospital. Nurses scan the bar codes of the medicines they are to administer and then scan the bar code on the wristband of the patient that is to receive the medicines. The system verifies that the right patient is receiving the right drugs and notifies the nurse if a potential error has occurred. This is a practical safety measure that, when executed properly, eliminates medication errors within the hospital. However, practice does not always follow official process. In some hospitals, it is common for the medication cart to remain in the hallway while the nurse brings the drugs to the patient. In these instances, the nurse scans the drugs at the table in the hallway and then scans the bar code on the patient's chart that is also in the hallway. Although this user-modified process greatly decreases the likelihood of errors (compared to not using any bar code system at all), the process still leaves room for error, making the system less effective than it could be at catching human error.

Although not all process alterations have such potentially deadly consequences, this example illustrates that human habits are hard to deny and even harder to change. As you work through your pilot and production deployments, evaluate whether operators use the system as your RFID team has designed it. Some user-initiated processes cause a decrease in performance, whereas other user-initiated process alterations yield more efficient operations. You may be able to redesign the system to overcome the problem in the process and take advantage of efficiencies that operators develop. After all, RFID systems are supposed to increase the efficiency of their work, not increase their workload.

When you can't accommodate user-initiated process alterations by redesigning the system, you need to retrain the operators. Remember to include in this training the reasons why the official process improves efficiency. Also consider making adherence to the official processes one of the performance metrics for performance evaluations. Thus, following the official process becomes a job requirement.

Explaining how RFID affects employees

Many employees and employee groups, such as worker's unions, are afraid of the impact RFID will have on jobs. And rightly so — a well-designed RFID network can eliminate some human jobs (like orienting packages so the bar codes are correctly positioned on a conveyor).

It's up to company leadership and the RFID committee to make sure that the workers affected by the RFID deployment understand exactly what is going to happen to their jobs and how they benefit from the RFID deployment.

Chapter 14

Ping-pong, the Tags Are Gone: How to Monitor Your RFID Network

I remember as a kid growing up, my mom would make cookies and she'd let me have a few when I got home from school and maybe a couple after I ate all my dinner, but that was it. Later, while Mom was watching TV, I would sneak into the kitchen and climb up onto the counter to reach those cookies on top of the fridge and add to my caloric energy store for a good night's sleep. What do cookies have to do with RFID? Well, not that much, except for the need to monitor them. Mom just wasn't very good at it, and, consequently, Pops didn't get any cookies by the time he got home. I don't want you to be like that with your RFID system because if you're not very good at monitoring it, Pops won't pay the price. Your bottom line will.

This chapter gets you up to speed on how to keep an eye on all your RFID readers. I explain the two basic types of monitoring. The first is simply checking that your reader is active. The other type of monitoring focuses on the behavior of your system. The goal of this type of monitoring is to devise ways of measuring how readers are working and then, using these measurements, detect, anticipate, and solve problems that arise with individual as well as multiple readers at the site. In this chapter, you find out how to set up monitoring as you roll out your RFID system and how monitoring helps you keep that system (and your bottom line) in good shape.

Why Monitor an RFID Station?

In a phrase — 24/7 — that's what you need out of most RFID systems once they are production. In the process of setting up an RFID system, you examine the physics of RFID and how various system components work together in your environment. This analysis ensures that the antennas in an RFID station are optimally positioned and aligned. A fully operational RFID system is only a first step, however. After you have selected, tested, and installed an RFID reader with its antenna array (the "station"), a relatively large number of problems can arise. Here are some examples:

- A reader can be turned off for maintenance or simply because it is not being used, and then be overlooked and remain off.

- A power failure can reset or wipe out the configuration of the reader.

- The antennas can be placed in a high-traffic area so that people or machines (such as forklifts) regularly block their ability to read tags.

- A reader can fail to detect tags or begin to detect ghost or spurious tags due to a hardware or software fault.

- The communication network between the reader and the middleware (savant) can go down or become overloaded.

- The antennas can be improperly placed (they may have worked well for your test cases, but fail to pick up tags in a production environment).

- Your site can be periodically (or constantly) flooded with radio frequency traffic in the same bandwidth as your RFID system (normally this would be 902 to 928 MHz). Hand-held barcode readers, as an example, often collide with the RFID spectrum.

- A change in the structure of your facility (such as adding new walls, building a new containment cage, adding or moving a conveyor, or installing a mobile wrapping station) can interfere with one or more previously working RFID stations.

- The reader or one or more of its antennas may be defective. Equipment defects often show up gradually over time.

- A reader software update by the vendor or new generation of tags can cause unexpected problems in reading tags.

- A reader can be burned out when the antenna leads are disconnected and power is still on.

- The strength of the antenna's electromagnetic (EM) field can be insufficient to read outlier tags or can be so strong that it reads tags in a neighboring RFID station (such as an adjoining bay door).

- Packages, pallets, containers, or other RFID tagged objects move too quickly to be accurately read.

These problems arise in the ordinary operation of an RFID workstation, and some of them arise on a more or less regular basis. Because every RFID station encounters performance and operational errors at some point, you must pay close attention to how the readers are working and know when any reader shows some sign of anomalous behavior.

Setting Up Two Types of Monitoring

Monitoring systems come in two basic forms, and you need both to keep your system running as smoothly as possible. The first form is basic status monitoring, and the second is monitoring the behavior of your system.

- **Monitoring the basic status:** This monitoring is your first line of defense in troubleshooting and preventing problems with your data. It includes having a light or other simple feedback system in place so that operators know whether a read is successful or unsuccessful and how to respond to unsuccessful reads. In addition, the system administrator monitors the basic status of the RFID network as a whole. Simple status-indicator panels on the administrator's computer desktop enable him or her to monitor all read locations from a single location. The administrator needs basic status information such as network connectivity, power to the units, and which antennas are operating.

- **Monitoring system behavior:** You monitor system behavior by gathering usage statistics, such as read rates and read accuracy. This data about the system enables the administrator to understand how the system behaves normally and to detect signs that something is awry. Particularly when plotted over time, both sudden and gradual performance decreases can be quickly identified.

Checking That a Reader Is Active

When a monitoring system first starts up, and also regularly throughout the day, it must interrogate each reader, essentially asking the reader whether or not it is alive. If the reader does not respond, it has been turned off, is unavailable through the network, or has suffered a fault in its internal software. This section helps you weigh your options for checking readers over the network. You also need to set up simple feedback for operators so that they can effectively monitor whether the system is working in order to get their jobs done.

Determining whether a reader is active is not associated with periodic behavior over time (like monitoring behavior). Although you don't normally expect to see a pattern of unavailable readers, patterns of unavailability are often critically important when they emerge. Such patterns often mean that a reader is suffering from a mechanical defect, that a site suffers from regular power outages, or that the power cables for a reader are subject to damage or accidental disconnection.

Choosing the right method

How do you determine if a reader is active? There are basically two ways. The first is through the `ping` command, and the second is through the network management capabilities of SNMP (Simple Network Management Protocol).

Although the `ping` command is the simplest method, it has some serious drawbacks. Foremost among them, `ping` cannot tell you whether the reader is actually working. The `ping` command is equivalent to calling your friend George on the telephone. If the telephone rings, you know that George's telephone is connected to the communications network, but you don't know if George is actually home or if the telephone handset is working. If George picks up the phone and says "Hello," well, then you know that the telephone is connected, George is home, and the telephone is able to sustain a communication link between yourself and George. An equivalent approach is needed for RFID readers. You need to know that the reader is connected (that is, it is physically visible on the network) and that it is capable of communicating with the outside world (that is, it is able to send messages; essentially, that someone is home inside the reader!). To do this, turn to SNMP.

SNMP was introduced in 1988 and has evolved as the standard for network management. Its ubiquity across nearly all families of network devices is due to its relative simplicity and its sparse code requirements.

The Simple Network Management Protocol (SNMP) provides a system as well as applications layer service protocol that allows the easy exchange of status and performance information between networked devices. SNMP is a component of the Transmission Control Protocol/Internet Protocol (or TCP/IP). By using SNMP, a system administrator (or a monitoring software system) can determine whether a device is visible on the network *and* query the device. By querying the device, an SNMP interface can determine not only that the device is connected, but also that it can send and receive.

A simple human interface: Enabling operators to monitor the system

As long as we humans are running and monitoring RFID systems, we must know how well the system is operating. Early detection of missed reads and faulty or ill-functioning readers reduce operating costs and make the system run smoothly. If you deploy an RFID network in warehouses, production floors, and the like, workers who use the system in the course of doing their jobs become your first line of defense in detecting problems with reads. Because these workers may have little or no technical knowledge (and also because you want to make their jobs simpler, not more complicated), you need a simple interface that enables these workers to monitor whether the system is working as it should.

The two simplest forms of feedback are sounds (beeps) and status lights.

- ✔ **Sounds** are very good for giving feedback on human-initiated actions, such as reading a tag with a hand-held reader. Simple beeps provide immediate feedback without the need for the operators to remove their eyes from the task at hand. Fixed-location readers, in which the operator actively monitors the activities, may also use sound as a feasible feedback. However, care must be taken not to overload the operator. A beep indicating the read of every object is easily tuned out and ignored when large numbers of objects are being identified.

- ✔ **Status color highlights (or simply a "color light")** are very good for providing feedback when the operator is looking for it or when he may otherwise be easily alerted by the presence of an alert color. When only discrete read locations are used, a status light indicating that the read location is in operation enables the operators to effortlessly determine that they either should or should not be receiving proper reads from the system. The status light on the reader equipment is not sufficient feedback because the readers could be located in protected areas that may obscure or block their tiny status lights. Also, multiple readers may be used to cover a particular read zone. Having the operator check each reader is not effortless and is error-prone. Simple status lights can let the operator know quickly whether all the readers in the area are working properly.

You have many options for attaching light or sound enunciators to readers. Most readers have several dedicated I/O ports designed specifically for adding motion sensors, lights, horns, and so on.

If you decide to use lights, include a single status light for each read zone. Make it easy to see with peripheral vision and easy to interpret. The mental image you should have for this status light is the large "X-Ray in Use" light and sign commonly seen in hospitals. When the light is on, it is hard to miss, and the sign tells you exactly what the light indicates. Make your status lights this obvious. Table 14-1 lists some common status light systems that help operators do basic monitoring and thus use the system effectively.

Table 14-1	Status Light Setups	
Type of Light	*How It's Used*	*Example*
Flashing lights	Attracting attention; alerting operators when a process running in the background stops working	A logistics and operations program may tell incoming trailers which dock door to use. If the RFID system at a dock door is having trouble, the operating program takes that dock door out of the scheduling queue and signals a red light to flash so that no one uses that door by mistake.
Yellow lights	Warns that some process is entering a malfunction zone or that the system predicts a system fault will occur	A reader begins to recognize fewer and fewer reads. This can indicate a fault in the equipment. When the number of valid reads begins to decay, the monitor predicts the rate of decay and turns the reader status to yellow, indicating that a working reader is about to fail.
Red light, green light	Checking that reads are successful and controlling the flow of goods where a limited number of tags or a specific type of tag, such as a pallet load tag, are to be read	On a conveyor belt, a green light means a successful tag read; a red light means a failed tag read. The conveyor may divert an object containing an unread tag or signal the operator with the red light that some other action is necessary. At a dock door, if a tag is read, a green light flashes, if someone goes through the portal and a tag isn't read, the light flashes red.

Several RFID racks on the market can be used between adjacent dock doors, with antennas on both sides of the rack. In these configurations, any light notification you set up should be clearly associated with its corresponding dock door. Some racks have lights on the top of the rack, so a dock worker at either dock door might see it flash and mistakenly think the reader is working on his or her side.

Measuring and Interpreting System Behavior

Although determining whether a reader is visible and working is an important factor in monitoring the health of an RFID system, it is not a sufficient measure of robustness and relative long-term stability. In order to gain a better understanding of how a station or a collection of stations is performing, you need insight into a station's behavior.

Behavior analysis relies principally on a statistical analysis of a reader's past performance and an estimate of the reader's short-term future behavior. When you understand how a reader performs, you can begin to detect unusual changes in this behavior, predict the failure of a reader, and determine the factors that might contribute to a loss in reader performance.

Although a wide variety of behavior measurements are available, this section focuses on a small handful that are sufficient to build a reasonably efficient monitoring system. They tell you when your readers are on line, when they are working properly, when they are affected by some environmental stress, when they are exhibiting abnormal behaviors, and when some emerging condition exists that will cause a problem in the near future. In addition to helping you find faulty equipment, these measures can help you evaluate vendors, recognize systemic equipment and network design or placement problems, and alert you to emerging problems as your RFID system "settles in."

Building a statistical monitoring approach

When monitoring an RFID system's behavior, you need information from the readers, which you can get through both nonintrusive monitoring and intrusive monitoring:

- ✔ **Nonintrusive monitoring:** Nonintrusive techniques analyze the information that is normally available from a reader and place no additional demands on the RFID system.

- ✔ **Intrusive monitoring:** You ask the reader to periodically provide information that is not available by simply interrogating the tags. You do this by issuing commands to the reader to tell something about its internal operations.

In nonintrusive monitoring, you might check the status of a reader's tag detection. This status condition can be successful (the contents were recognized) or unsuccessful (the response to a reader's attempt to interrogate a tag was not recognized). Intrusive monitoring can often give deeper insight into an RFID system by providing a way of not only predicting failures, but also helping to establish a root cause analysis that can bring a failed reader back on line in the shortest possible time. As an example, the ratio of successful to unsuccessful responses over time is an indicator of how well the reader is positioned or, following a decay curve, that the reader is suffering an imminent hardware failure.

Deciding whether to use intrusive or nonintrusive techniques usually depends on three factors:

- **Availability of programming resources:** Nonintrusive measurements are often easy to come by. The Savant, or middleware systems, attached to most readers usually provide counts of unique tag values and other statistics, which can be retrieved using simple programming techniques or, in some cases, are available as files without any additional programming. In order to measure many of the metrics in an RFID reader, you may need the services of a computer programmer skilled in embedded systems and network protocols. If these skills are not available in your organization, you need a budget to hire programmers with these skills.

- **Availability of a comprehensive command language for the reader:** An RFID reader is a small computer in its own right. The actions of the reader are controlled through its internal programs, which are written in its own programming language. The reader's program language often includes a series of commands that you can use to retrieve information about how the reader is working. This is a "command language." Many reader command languages are very close to the basic machine instructions and involve cryptic syntax with brittle parameter lists.

  Accessing information about how the reader is working and what is happening at the bit level is only available through intrusive monitoring techniques. This also means that using a reader's command language often requires highly specialized computer programming knowledge. Command language programmers also run the risk of accidentally disabling the reader, corrupting or destroying configuration data, or causing the reader to act in an unpredictable manner or to return the wrong data.

  Some more advanced readers that are connected through TCP/IP or Ethernet networks strike a happy medium between the two extremes by using a Hypertext Transfer Protocol (HTTP) server (such as Jetty) as a way of communicating with the reader. An HTTP server (though still requiring professional programming expertise) interrogates and communicates with the reader through a simple set of easy-to-use-and-understand commands. In general, an HTTP interface simply taps into data that is already available form the reader's operation. In this sense, HTTP monitoring tends to be a form of nonintrusive monitoring.

> ✔ **The load on the reader during the busiest time of day:** An intrusive monitoring approach places a demand on the available computing capacity of the reader. If this demand is too heavy, it can affect the ability of the reader to work properly under a heavy load. From experience, this doesn't happen often, and sometimes the valuable operating and performance information derived from intrusive monitoring outweighs the small probability that a rare spike in tag traffic will cause a problem. On the other hand, if you have periods of very intense tag traffic, you should consider approaches to nonintrusive monitoring.

In general, the statistical analysis, trend analysis, and visual representations of nonintrusive values provide a wealth of monitoring information for a basic RFID system. In many cases, you can also learn quite a lot from the simple aggregation (summation) of tag counts and tag count failures by location, by time of day, or by reader manufacturer type. From these statistics, the operations manager and the system user can gain a significant insight into how the readers are working and where problems occur or where they are likely to occur in the future.

Starting with data that's easy to extract and use is the best approach to building your monitoring system. As you gain experience and discover any weaknesses in your monitoring approach, you can plan for a more in-depth analysis. As you consider additional monitoring, remember that monitoring should be goal-oriented. Collecting statistics and plotting reader transmissions without a method of understanding what is happening in your enterprise is likely to result in a static view of your equipment. A monitoring system must tell you what is happening and, to some degree, what is likely to happen.

As a first step in developing your monitoring strategy, investigate the command language supplied with your reader and decide whether the added investment in programming and the possible added load on the reader is worth the investment in accessing additional operating information. You should also consider whether a commercially available (off the shelf) RFID monitoring system would best suit your business needs. These systems are designed for minimum impact on your readers and can often provide a wide spectrum of performance statistics.

Breaking data into time intervals

Before examining performance measurements (which are the metrics you use to evaluate how well a station — or a set of stations — works), you must understand a simple but often overlooked property of these measurements: Measurements have different values and hence different meanings at different times. Generally, a given measurement might have one value in the morning and another at night, or one value on Monday and another on Wednesday, or one value during the beginning of the month and another toward the end of the month. The variation of measurement values over time is a result of the recurring or cyclical nature of a behavior. Somewhat akin to the seasonal changes in temperature, unemployment, or produce pricing models, the regular periodic behavior of an RFID station means that you need to consider *when* as well as *what* when measuring performance.

In addition to the operational performance measurements discussed in this chapter, you should also provide your monitoring system with some solid statistical analysis capabilities. Understanding the nature of the data provided by a reader is critical for interpreting and effectively using the results of each performance measurement. When you have captured a collection of tag instances with their date and time values, you can begin to answer some fundamental questions about the organization and distribution of the data. To begin a statistical analysis, you need to group the data into small "buckets" or intervals based on the time values. In this way, you have a count of tag reads every two minutes (as an example). Eventually you will create a table of two-minute counts over a few hours or days. As an example, Figure 14-1 illustrates how a collection on frequency counts for a reader at a particular site can be tabulated.

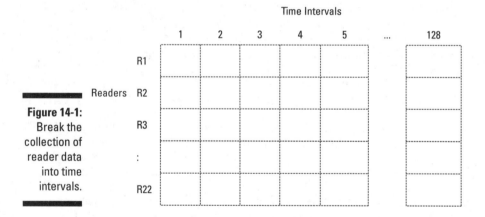

Figure 14-1:
Break the collection of reader data into time intervals.

This is where statistics come into play. Some of the fundamental or descriptive statistical measures you want to develop include the minimum and maximum frequency counts, the average value, the standard deviation of the data distribution (which is a measure of how compactly the data is gathered around the mean), and the skew of the data (which indicates whether the data has a bias, right or left, away from the mean). And, although you need to compute these measures for all the data in order to give yourself some sense of the overall data properties, you want to use the time stamps to break down statistics into more meaningful time frames, such as morning and afternoon, for each day of the week. This emphasis on time, of course, is in keeping with the fundamental time property of all your measurements.

Please bear in mind that not every manufacturer's RFID reader provides all the information necessary to compute every measure. In some cases, you need to use intrusive programming techniques to retrieve a required metric (such as a failed read count). Also, in some cases, you may need to add the time of day measure to a metric that is returned without a valid time stamp.

Measure 1: The average tag traffic volume (ATTV)

This measurement enables you to understand the typical flow of tags through the station and anticipate the average volume of tags that pass through a station in a given interval of time. To capture this flow, your monitoring system captures the number of tags per time period (as an example, tags per minute or tags per every five minutes). To compute this measurement, you need to gather the following data from your readers:

- **The time those tags were counted,** which you then use to divide tag counts into a time interval that you specify for the statistical analysis of ATTV
- **The tag count coming from the reader,** which you may need to limit, depending on how your readers work (I explain this in more detail a little later in this section)

You want your monitoring system to collect enough data so that you get a feel for the station's tag volume behavior through the week. With this day-by-day, 24-hour volume data in hand, you can use a spreadsheet program such as Microsoft Excel to compute trends, discover periodic behaviors, and graph the flow of tags through the station. I discuss how you can use ATTV as a foundation for more complex monitoring later in this section.

Because the number of tags read in each time interval can be very large, the time interval determines the level of granularity you want to use in evaluating this measurement.

If the time interval is too large, the final measurement doesn't tell you very much because too many reads over a large period of time obscure the patterns of tag flow. On the other hand, if the time interval is too small, the fine details similarly don't allow you to see the patterns of tag flow.

Finding a good value for the time interval is generally not difficult, but does require a bit of trial-and-error testing. Generally, you know you've found the proper time interval when the patterns of reader activity begin to "make sense." A good starting point is five to ten times the sampling rate. Thus, if you're sampling every two minutes, a time interval of ten minutes would be a good starting point.

Tallying the tag count with your readers in mind

Although the tag count appears simple, your monitoring system needs to account for the workings of the RFID reader itself. For example, as a package moves on a conveyor, the transponder tag is activated and transmits its EPC (electronic product code) value. But it doesn't necessarily do this only once; it might conceivably do this over and over again as it moves along the conveyor. Your monitoring system needs to be configured so that your monitoring isn't thrown off if the conveyor stops.

- ✔ **For more sophisticated readers that send state changes along with a tag's information,** you can calculate ATTV using a standard equation that doesn't need to be revised to accommodate your reader's behavior. After a tag is read, no more EPC values are sent even if the tag sits in front of a reader for a long time. When the tag starts moving again and leaves the reader's detection zone, the reader sends a message telling the listener (the middleware, as an example) that a state change has occurred — the tag has moved out of range.

- ✔ **For readers that continuously transmit tag values (or for readers that are continuously interrogated by the Savant software, which results in pretty much the same situation),** the way tags move through the interrogation zone influences the average traffic volume measurement. Although spikes, plateaus, and deep valleys in the tag count might conceivably reveal much about the way material flows through your warehouse or retail store, ignoring these irregularities in traffic has a significant impact on the use of the traffic volume as a reliable metric, and you need to account for it.

One easy way around this situation is to simply treat each time interval as a bucket. When filled, all other tag counts are thrown away. When you tally tag reads this way, you need to set a maximum number of counts for the specified time period. As long as the total tag count is less than or equal to the maximum number, your data reflects the actual number of tags that move through the interrogation zone. When the tag count hits the maximum number, the monitoring system stops counting.

Because the purpose is to develop a way of monitoring the activity of a reader, the measurement need not be absolutely precise and analytical, as long as it is consistent and reliable. So setting a maximum number of counts doesn't diminish the value of your monitoring system. One way to initially begin using the bucket approach is to set the maximum count to about 80 percent of the estimated maximum possible during a single sampling period.

Using ATTV to monitor whether a reader is active

The average volume measurement can also be interpreted as a measure of whether a reader was actively reading tags during any time period. In this case, you only need to know that at least one tag was interrogated. If you use the bucket method and set the maximum number to one tag, the equation in your monitoring system produces a series of 1s and 0s over the time period.

The 1s and 0s produce a Boolean array indicating whether or not a reader was active during the time period.

Note that this approach differs in purpose form the use of SNMP to determine whether or not a reader is active. The SNMP will tell you whether or not the reader is visible on the network and is working as far as the network of readers is concerned. The average volume per time unit measures the functional or operational activity of a reader – that is, whether or not a reader is capable of reading tags (a reader with a damaged or disconnected antenna will still appear active to SNMP interrogation).

Measure 2: Read errors to total reads (RETR)

An *error* is a failure to interpret a returned signal from a transponder tag. By measuring the errors against the total reads, you can see how well an RFID station scans and recognizes tags and thus detect problems, such as

- ✔ A faulty tag, antenna, or network connection
- ✔ Improper placement of antennas
- ✔ Improper tag types for the kind of container material
- ✔ Signal interference in the range of RFID frequencies
- ✔ A low signal strength
- ✔ A frequency hop time that is longer than the tag time through the antenna's detection field

Read errors are associated with the number of times that an antenna must probe the incoming container to properly detect the tag. To compute this measurement, you need to collect the following data from your readers:

- ✔ The number of read errors
- ✔ The number of successful reads
- ✔ The time these reads occurred, so that you can evaluate the measurement over different time periods

The error rate measure is the number of read errors over the time interval (N) divided by the total number of reads (the number of successful reads plus the number of read errors) in this time period. In effect, this is the percentage of errors in a given time period. A high percentage of errors is cause for concern and investigation.

Measure 3: Read error change rates (RECR)

In addition to the read error rate (which I discuss in the preceding section), you need to know the change in this error rate. The rate or degree of change in the error rate measures the stability of the RFID station.

If the read error rate fluctuates or increases steadily, some underlying problem must be interfering with the reader performance. These underlying problems often emerge over time and reveal a fault in either the design of the station or in the hardware itself.

Two different types of methods can detect changes in the error rate — essentially, a simple method that detects a *degree of change* and a more complex method that detects the *rate of change:*

- **Degree of change:** This simple method looks at the cumulative difference in the RETR values over time and answers the question, Are the values, on average, steadily increasing or decreasing?

- **Rate of change:** Without delving into the statistics, this method may bring back memories of plotting graphs in algebra class. This measure involves plotting the values on x- and y-axes (in which x is the time interval and y is the RETR value), finding a trend line through the values, and then determining the slope of the trend line. With this measure, you know whether the RETR values are increasing or decreasing *and* the rate at which the change is moving up or down.

Measure 4: Actual versus predicted traffic rate (APTR)

Earlier in this chapter, I discuss the average tag traffic volume detected by a reader across various time frames. (See "Measure 1: The average tag traffic volume" for details.) As you collect this traffic information, various patterns begin to emerge. For example:

- Warehouse bay door readers in your distribution centers have more or less predictable tag volumes that change on different days of the week or months of the year or even times of the day.

- RFID inventory tracking systems in retail stores often have even more clearly defined periodic behaviors, corresponding to store hours, after-work and weekend customer traffic patterns, and so forth.

From these historical observations, you can begin to detect anomalies in the performance characteristics of your readers. Your system may use one of two approaches to measure APTR:

✔ **The simple method:** One approach is simply to compute the ratio of the average tags per interval against the actual tags per interval rate and raise an alarm if the ratio is smaller or larger than expected.

✔ **A more complex and slightly better method:** I say better because this method helps you both detect anomalies as well as predict problems. In this method, your system forecasts the expected tag traffic over a short-term period and then compares the actual to the forecast. For this method, you need the following data:

  • **The time tags are read,** which you then use to compare actual reads in a time interval to reads forecasted for that time interval. The time period is the time of day usually expressed in minutes since midnight. This creates a continuous, increasing value that can be used in the prediction process.

  • **The actual ATTV value** for the reader during the forecasted time period.

When you examine the data for this measure, an actual value that's close to the estimated value indicates that the system is running fine. If the actual traffic is much larger or smaller than the expected traffic, you need to investigate. First, determine whether your actual volume has changed — that is, whether the change reflects what's actually happening in your business. Then, investigate whether the change reflects a problem with your system. Both of these measures might be important, but a smaller than expected outcome often indicates some problem with the reader.

Measure 5: Mean time between failure (MTBF)

The measures I explain earlier in this chapter are operational in nature; they form the basic metrics that you can use to detect problems in your RFID system as it is running in your organization. The *mean time between failure* measure, however, is a conventional engineering metric and is part of a strategic or global measure that measures the robustness and stability of a system.

MTBF is a high-level or strategic measure because it's a measure of system reliability and is designed to assess the performance of a large system of "things." In your case, these "things" are the components of your RFID system — the readers, the antennas, and the underlying network. When you apply this measurement to an RFID system, you measure how well this system of components performs over time given its site deployment, ambient or environmental threats, and the continuous flow of tag traffic.

When you measure MTBF, you need to determine what a failure is exactly. How you define a failure in turn determines how sensitive your monitoring system is. You can basically steer your monitoring system toward one of the following extremes:

- ✔ **A very hard approach** defines failure as a reader that is not operational.

- ✔ **A softer approach** defines failure as a decay in one or more performance measurements below a certain threshold. Raising or lowering the threshold provides a way of identifying problems based on varying ideas of criticality or sustainability.

Generally speaking, the softer approach is a better way of looking at failure. Because a reader "fails" when its performance crosses a threshold, it provides a form of early warning. This early warning is often sufficient to bring a defective reader or network component back on line.

MTBF measures the expected failure rate over time based on the failures rates of the individual components. To calculate MTBF, you need to determine or gather the following data:

- ✔ **The lifetime:** This is the total functioning life of the population and is measured as the activities in a time period times the length of the time period. The population is the total number of readers that you are monitoring or evaluating. Activities are the number of reads associated with the readers over a specified time period. The lifetime is usually measured in hours.

- ✔ **The number of components in your RFID system:** Usually, when measuring MTBF, a component is a reader. However, a component might also be an antenna or a network server. Taken together (readers plus antennas plus servers), these comprise the components in the system. Each of these might have a failure rate.

- ✔ **The quantity of *each* component:** If the system consists of readers and antennas, this is the number of readers and then the number of antennas.

With this information, the failure rate is associated with that component. The MTBF rate is a measure of how many failures that component has had over the measured life of the system.

What makes MTBF so interesting and so potentially important to a monitoring system is its ability to fuse together the failure rates of a system with many different components. Because an RFID system has many components — readers, antenna arrays, networks, and middleware or Savant servers — the MTBF statistics provide a powerful way of predicting failure rates based on the "weakest link."

Here's a simple example to help explain how to determine the variables for and to calculate MTBF. Consider a calculation of MTBF solely for RFID readers. The readers are the only components in the system. Suppose you have a system with a total of 1,000 RFID readers operating over a 100-hour period, and during this time, 10 readers fail. The lifetime is the 1,000 readers times 100 hours — or 100,000. You have only a single component (the collection of readers), and its failure rate during this time period was 10 units. If you plug these values into a basic statistical equation for MTBF, this works out as follows:

$$mtbf = \frac{1000 \times 100}{\sum_{i=1}^{1} 1 \times 10} = \frac{100000}{10} = 10000 = 10^4$$

The units of measure in this case are hours divided by elements (or units) so that the MTBF for a population of RFID readers is 10,000 hours per unit. If you divide those hours by 24, you get an idea of how many days you can expect your RFID system to run between failures — in this case, 416 days. Of course, MTBF is a statistical measure that applies to a large population and is only a rough measure of reliability. Like all statistical analyses, its applicability to a particular RFID system depends on the accuracy and quantity of the underlying data for each component in the system.

Monitoring as you expand your RFID network

The physics of an RFID system and the techniques used to develop a preliminary site assessment (see Chapter 7) provide a proven and controlled method of placing reader stations and antennas. In practice, however, expect that subtle changes in the operating environment, trade-offs between physics and practicality, and the ambient signal capabilities in a working environment will have an appreciable and noticeable effect on the actual capabilities of your system.

The monitoring system plays an important role when an RFID station is reconfigured and redesigned because it provides rapid feedback on a reader's new success or failure rate in reading tags. Fine tuning your RFID system with a working monitoring system enables you to quickly isolate problem areas, take advantage of favorable antenna locations, and measure each increase in performance.

Setting up a monitoring system

After you understand what metrics you need to measure, you're ready to set up a monitoring system. You want to make sure that this monitoring system is built based on the statistical equations in Table 14-2. Although explaining the details of how to calculate statistical equations is beyond the scope of this book, I do explain the basics of statistical notation in the sidebar, "Some statistics basics," and the preceding sections explain what these equations calculate in basic terms.

I include these equations so that you can clarify how you want your programmers to build a monitoring system or how you want your monitoring vendor to measure your system. But more on that in a moment. . . .

Some statistics basics

A full treatment of statistical methods is beyond the scope of this book. If you are interested in brushing up on your statistics, the Rice Virtual Lab offers some basic introductory information (www.ruf.rice.edu/~lane/rvls.html).

In you addition, you may want to understand one of the most frequently used symbols in statistics, the sigma or summation symbol, because you encounter this symbol in most of the statistical-based measurements in this chapter.

$$\sum_{i=1}^{N} x_i$$

The summation notation in this example is read, "From 1 to N, sum the value of x". In this example, N indicates the number of elements in the distribution, i is a counter or specifier for each value in the distribution, and x_i is one of the values in the distribution. If we have a distribution {1, 2, 3, 4, 5}, this expression selects each value and adds it to the ongoing sum. The result is 15.

Table 14-2	Monitoring Equations	
Measure	**Equation**	**Variables You Supply**
Average tag traffic volume	$m_1^P = \dfrac{1}{S}\sum_{j=0}^{S}\left(\dfrac{1}{N}\sum_{i=0}^{N}t^{(i)}\right); P = \{1, T\}$	t = a tag detected by a reader N = time interval S = the number of instances of N you want to measure, to break the results into time periods and look for patterns T = number of hours (set this to 24 to see patterns during a day) P = time periods
Average tag traffic volume, with a limit on the maximum number of reads to accommodate reader behavior ("the bucket method")	$m_1 = \dfrac{1}{N}\sum_{i=0}^{N}\min\left(t^{(i)}, B\right); t^{(i)} \le B$	t = a tag detected by a reader N = time interval B = bucket size

Measure	Equation	Variables You Supply
Read errors to total read rate (RETR)	$$m_2 = \dfrac{\sum\limits_{i=1}^{N} e_i}{\sum\limits_{i=1}^{N} e_i + \sum\limits_{i=1}^{N} s_i}$$	N = time interval e = error of one type or another s = the number of instances of N you want to measure, to break the results into time periods and look for patterns
Degree of change in read error rates (RECR)	$$m_3 = \sum_{i=1}^{T} \left(m_2^N - m_2^{N-1} \right); N > 1$$	N = time interval T = number of hours (set this to 24 to see patterns during a day)
Rate of change in read error rates (this measure has three parts)		
Linear regression for trend line	$y_i = a + bt$	t = time period y = estimated value of RETR a = value for the y intercept b = slope of the line
Standard error of estimate	$$s_{est} = \sqrt{\dfrac{\sum\limits_{i=1}^{N}\left(y - \overline{y}\right)^2}{N}}$$	N = time interval y = estimated value of RETR
Rate of change read error rates	$$m_3 = \overline{s} = \dfrac{1}{N}\sum_{i=1}^{N} s_i$$	s = slope N = time interval
Actual versus predicted traffic (APTR)	$$m_4 = \dfrac{1}{N}\sum_{i=1}^{N}\left(y_e - y_a\right)$$	N = Number of time intervals in the future y_e = estimated value from regression equation y_a = actual value for reader during the time period
Mean time between failure	$$mtbf = \dfrac{l_t}{\sum\limits_{i=1}^{N} q_i \times f_i}$$	N = time interval l = lifetime t = tag detected by reader q = quantity of components (readers for example) f = failure rate associated with a component

Depending on your needs, the availability of programming resources, your budget, and the capabilities of current off-the-shelf monitoring software, you can use any of the following resources to build and monitor your system:

- **Programmers, either in-house or consultants, who build a custom monitoring system.**

- **Monitoring vendors, who are outsourcing partners and who supply results to you:** In this case, you need to tell your outsourcing partner what you want to measure, and you can use the equations in Table 14-2 to define the nature of the monitoring system. See Chapter 17 for more details on outsourcing.

- **Third-party software:** RFID is a new, emerging technology. As I write this book, only a few monitoring systems are available, and these are often associated with a single vendor. As the market expands, you should expect to find monitoring systems, often developed by middleware vendors, of various capabilities, sophistication, depth, and cost.

A self-healing system

Because human intervention can create errors and inconsistency, the ideal preservation for an RFID system is a monitoring system that can self-heal after it has detected anomalous behavior. I've also stressed this in the design of ODIN technologies monitoring system because there aren't a lot of technology people to fix things where warehouses tend to be — like Nome, Alaska or Waterloo, Iowa. The reader configuration classes are the basis for that type of self-healing system. (As I explain in Chapter 13, configuration classes are standard reader configurations that you can roll out to a reader over your RFID network.)

Intelligent software is emerging that will allow adjacent readers to configure each other based on performance and RF analysis. ODIN technologies applied for one of the first patents in 2003 for such an automated design system. As readers become more intelligent, the need for specialized testing equipment, like signal generators and spectrum analyzers, will decrease. As more sophisticated systems start to address on-going reader maintenance issues, first the machine will try to remediate any anomalous behavior, and then a notification will be sent out a system administrator.

Part V
How to Speak Bean Counter

"We can monitor our entire operation from one central location. We know what the 'Wax Lips' people are doing; we know what the 'Whoopee Cushion' people are doing; we know what the 'Fly-in-the-Ice Cube' people are doing. But we don't know what the 'Plastic Vomit' people are doing. We don't want to know what the 'Plastic Vomit' people are doing."

In this part . . .

*P*art V covers the suit-and-tie portion of this RFID business. It's written for the long-repressed bean counter in you. In this part, I help you figure out how to put together a working RFID committee, develop a strategy, and make a business case. I assume that you will need to go through an investigation phase, submit something during strategic planning cycles, and finally build a return-on-investment (ROI) analysis. This part takes you through all these steps and helps you decide whether to outsource.

This part may seem like it is just for senior managers and business analysts, but everyone can benefit from understanding how to get buy-in, how to plan the pilot program, and eventually how to deploy the RFID system.

Chapter 15

Making the Business Case

Cool new technology? Not cheap? Big strategic advantage? . . . Okay, I'll do it! Unfortunately, in most organizations, it's just not that easy. You have to contend with the bean-counting police before you can actually start having fun. That's why, in this chapter, I explain how to put together an analysis of this technology that any card-carrying CPA-, MBA-, CFO-type would be proud of.

A business case provides the overall business justification for the initial and on-going commitment of time, resources, and funding for an RFID implementation. One of the major lessons I've learned from large implementation efforts over the past 15 years is that the lack of a business case invariably leads to a lack of success. RFID is here, and failure is clearly not an option!

A critical success factor to your business case is your ability to execute with a rigorous and disciplined process. This chapter reveals a best-practice, nine-step process that will set you up for a successful RFID business case and clarify how to present it. For each of the nine steps, I define a purpose and tasks, followed by some experience-based how-to-get-it-done discussion.

Finding the First-Round Draft Picks for Your RFID Team

Prior to the project kickoff, you need to establish a core team of players who are already interested and motivated by what RFID can offer your business.

Look for people who are not only excited by the project, but who are also capable of conceptualizing how RFID can transform your business and who have the credibility within your organization to effectively evangelize the message.

Successful team members need to be able to push themselves and others to look far beyond the standard supply-chain-efficiency type of RFID benefits. A successful team also needs a structure:

- ✔ **Steering committee:** This is a small group of three to five executives who are the final budgetary decision-makers. The steering committee's role is to provide guidance and oversight to the project.

- ✔ **Project lead:** The project lead is a core team member who leads the business case project. In addition to having the right leadership and experience level, choose the project leader from the part of the business that has the most to gain from RFID deployment. Usually the project lead is someone in operations who will benefit from the RFID system or someone from IT who owns the system.

- ✔ **Core team members:** Like a great football team, you need players who are proficient in different disciplines — from kickers to linemen. For some companies, this step is akin to a puzzle wrapped up in an enigma, encircled by a cruel riddle. But it doesn't have to be that difficult — just use some common sense. Choose your key draft picks from finance, security, sales/brand management, information systems, and operations/logistics. You might have leaders who are already skilled in the topic or have some experience with the technology. You also need a technological guru who will be the primary leader for all things technological. This person's forte is the bits, bytes, chips, tags, hardware, and software — not the business processes. This person must understand the business, but doesn't need to be an expert. Understanding the business process is critical, however, so make sure that you have one or two members who really understand how things work on the manufacturing floor or in the warehouse.

Some companies are pressing for new structure here, such as a VP for RFID. Do you really need that? The likely answer is no, not unless RFID is going to be one of your core competencies. What you need is leadership, not a new office with new costs, without the organizational power and respect to make things happen.

- ✔ **Extended team members:** These individuals are normally brought in as required for specific steps or tasks over the course of the project. Typically, extended team members are key to the build-out of quantifiable benefits and costs. An example of an extended team member is someone from your customer service department who can provide the kind of detailed data collection and analysis that's necessary to establish credible benefits in the area of returns management.

Figure 15-1 shows a typical team structure, with the core team shaded in dark gray.

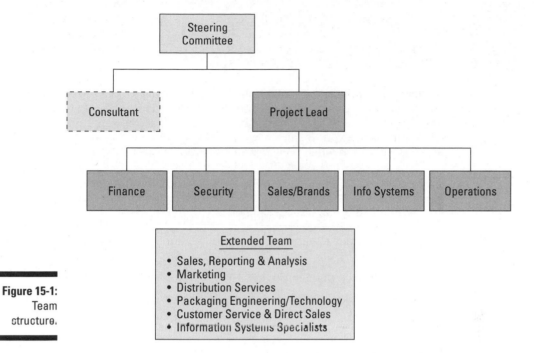

Figure 15-1:
Team
structure.

Many companies employ a consultant to lead them through the business case effort, typically because no one in the company has the proper RFID and business case process expertise and can be dedicated full-time to the RFID business case effort. If, like most companies, you bring in a consultant, select wisely. Make sure the consultant has experience doing RFID business cases in your industry. Be sure to review examples of his or her work as part of the selection process. If you look to some of the early adopters in the technology — the Wal-Mart top 100, for instance — many of the RFID leaders for those companies have spoken with the press. Call or e-mail those folks and ask whom they have worked with. Talk to some of the technology analysts like Yankee Group, Forrester, IDC, ABI, or Gartner. Also look at consultants who have been involved for a while with associations like AIM Global, CompTIA, or the CEA.

Make your RFID implementation a company effort, not a departmental affair. Cohesion is the key to a solid company-wide RFID program. If you fragment your RFID effort, your implementation will fail. From every point of view, an RFID implementation revolutionizes and touches almost every area within the company. Because RFID has received so much press coverage, everyone wants to know about it. If your company has an often-visited intranet or portal, setting up a section specifically on RFID is a great way to keep everyone from sales to operations up to date on the progress. If you have a project timeline and milestones set up on the intranet, it's also a great way to publicly and collectively audit progress of the project.

A Game Plan Is More Than X's and O's — Use a Proven Methodology

By using the methodology I describe in the following sections, you can launch your company's RFID initiative effectively. This methodology establishes a complete RFID business case story — what you want to do, why, what it means to your business, what actions you need to take, what the investment is, and when the investment is required.

To give you an idea of how you might work through this methodology, Figure 15-2 shows the nine steps over a ten-week project life cycle. Note that this is an aggressive approach that assumes

- ✔ A core team is already in place, and its members can each dedicate six to nine hours per week to the business case.

- ✔ That a project lead or consultant is working on the business case full-time.

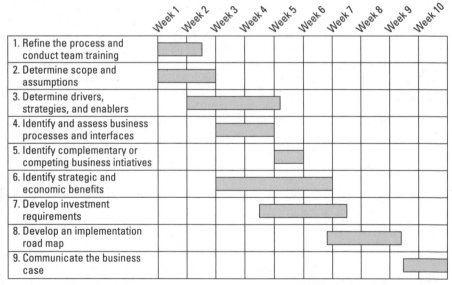

Figure 15-2: The nine-step approach.

Part of the methodology I explain in this chapter is based on building out an Excel-based ROI (return on investment) model. For benefit and cost estimates, coming up with low and high estimates is important. The variables for the low and high estimates should be based on the specific nature of the method used to determine the benefit or cost, be it time, assumptions on range of costs, or something else. While the various calculations and graphs produced by the ROI model are vital to the business case, the strategic bene-fits side of any RFID business case will prove to be the basis of any key decision-making. In this chapter, you find out at what points your team needs to identify strategic benefits and build elements of the ROI. In Chapter 16, you can find more details about strategic benefits and actually calculating the ROI analysis.

To get the best results, the individuals on the RFID team must function like a single unit. Efficient communication is the linchpin of an effective team, which is why I recommend two types of meetings:

- ✔ A weekly two-hour core team session focused on project status, issues, and collective review of work products (also called *validation*). Have this type of meeting as you move through each step of compiling your business case for RFID.

- ✔ An as-needed working session with core and extended team members focused on specific tasks.

When planning these meetings, take into consideration that some steps I cover in this chapter are more challenging and time-intensive than others (refer to Figure 15-2). These steps typically require several iterations during the weekly core team meetings in order to revise and finalize the work prod-ucts and findings (in other words, the steps in the nine-step process are not linear). Also, some require the full effort of the team, whereas others require one or two people acting on their own. When the project is aggressively pur-sued, it should take ten weeks.

Step 1. Refine the process and conduct team training

This step creates a solid foundation that enables everyone on your team to work successfully through the rest of the steps. To create this foundation, your team needs to accomplish the following tasks:

- ✔ **Train the core and extended project team members.** Training is pro-vided by the consultant and consists of RFID training and business case process training.

  - • **RFID training** consists of an overview of RFID tags, hardware, soft-ware, and case studies for the core and extended project team members. (See Chapter 5 for an overview on RFID readers and tags; Chapter 6 includes case studies of common RFID applica-tions). For technical team members, training also needs to include an overview of RFID technology (see Part III).

- **Business case process training** serves to educate team members about the overall business case process and then delves into business process concepts, issues/risk measurement, and best-practice project management.

To help you get up to speed on the technology, research RFID online (Chapter 19 lists sites that can get you started) and attend an educational seminar. Preferably find a seminar that is vendor-neutral or a conference with educational tracts and case studies. These seminars are real eye-openers and generally cover a smattering of all RFID-related topics.

- ✔ **Identify participants and roles for each step in the process.** Determine who is needed, how much of their time will be necessary, and when.

- ✔ **Refine the process for the RFID business case and establish an agreed-upon framework for an ROI model.** For details on setting up and working through an ROI model, see Chapter 16.

- ✔ **Review and finalize the objectives and deliverables.** Examples of objectives include establishing a recommended plan for RFID deployment and assessing the business risks of RFID deployment. Examples of deliverables include an executive level presentation and an implementation roadmap.

- ✔ **Establish consensus on the business case methodology and assumptions (I'm going to assume you use the 9-step method in this chapter).**

Before any meetings that involve people new to the project, send those folks an overview e-mail that introduces RFID basics, the purpose and scope of the project, and so on. An e-mail like this brings new extended team members up to speed quickly before they participate in any meetings or workshops.

Step 2. Determine scope and assumptions

Now your team is ready to set the big picture for the RFID implementation in the business case. To do so, here are the tasks that lay before you:

- ✔ Complete a scope statement that defines the RFID vision (the big picture of how RFID affects the business). The statement also identifies business units and processes included in the business case (or *in-scope*) and those not included (or *out of scope*).

- ✔ Establish operational assumptions for the business case.

Completing the scope statement

You need to establish your scope statement right up front. (Later on, you'll likely need to fine-tune the scope.) The key elements of an effective scope statement are as follows:

- ✔ **Geography:** Outline where the business case is focused (for example, North America, Asia). Is it intended to be a scalable business case to other geographies (that is, do you want to reuse the business case if you acquire another company in Europe, for example)?

✔ **Organizational coverage:** Identify what parts of the business the RFID will affect (for example, sales, marketing, manufacturing, distribution, information systems, finance, security).

✔ **Not included in organizational coverage:** List the parts of the business that an RFID system *won't* affect (for example, research and development, sourcing).

✔ **Products:** List the specific products that you want to tag. Normally, 20 percent of a company's products represent 80 percent of its sales. This is the 20 percent that you want to have within scope.

Establishing the assumptions

Normally, a business case has about a half-dozen assumptions. Assumptions are necessary to ensure a common understanding of how the RFID system will be implemented and what processes will be affected. Here are several examples of key assumptions:

✔ **All RFID tagging is done at the unit, case, and pallet level.**

You need to establish early in your business case what level of item you'll be tagging. For most retail consumer packaged-goods manufacturers, tagging occurs at the case and pallet level. For others, such as pharmaceutical manufacturers, you need to tag at the unit, case, and pallet level. At this point in the process, the level you determine is like a hypothesis. You'll prove or disprove your hypothesis during the alignment of drivers, strategies, and enablers in Step 3.

✔ **The business case planning horizon is five years.** Planning beyond five years is not viable with this rapidly changing technology.

✔ **The operational RFID flow begins at the point of tagging in manufacturing, goes to the manufacturing distribution centers, and then out to the trading partners.**

✔ **Tags are applied at the point of manufacturing or packaging (not slapped on in the distribution centers).**

Step 3. Determine drivers, strategies, and enablers

In this step, you identify and align the key drivers, strategies, and enablers. In the most basic terms, the driver tells you why, the strategy tells you how, and the enablers tell you what. Specifically,

✔ *Drivers* are external forces driving your business.

✔ *Strategies* are key RFID-related policies you'll use to address the drivers.

✔ *Enablers* are things you need to have or employ to achieve your strategic goals for RFID success.

All the business case benefits and costs (which you determine later, in Steps 7 and 8) emanate from aligned drivers, strategies, and enablers. Keep in mind that the better job you do in this step, the better grounded your business case will be.

Completing a strong set of aligned RFID drivers, strategies, and enablers is an iterative process. Here are the steps you need to work through in order to get this task done:

1. **Gather information about drivers, strategies, and enablers by reviewing strategic business plans and/or interviewing senior management.**

For key management interviews, be sure to prepare an interview guide with introductory material. In Part VI, you can find good introductory information about equipment vendors, general information on RFID, and RFID standards and protocols. In addition, an effective interview guide includes questions that elicit driver and strategy input and questions that help determine overall project expectations.

2. **Identify drivers based on your information.**

You can likely isolate four to six external forces that drive your RFID implementation. Drivers vary by industry, but most involve compliance, product safety or integrity, and, of course, your customer.

3. **Identify three to eight strategies for each driver.**

Some strategies are likely in place already. Your team can discover and compile those strategies through interviews with core team members and other stakeholders and then link the strategies to drivers.

Strategies start with action words or verbs. This will probably be the first time all your RFID-related strategies are articulated.

One of the overarching strategies is how much of an industry leader your company wants to be. This impacts the road map that you develop in Step 8, as well as how costs are incurred over time (Step 6). For more about formulating an RFID strategy from a business perspective, see Chapter 16.

4. **Identify the enablers that you need to have or employ for RFID success.**

Similar strategies usually have similar enablers, and you'll likely find a core of about a dozen enablers in your analysis. The statement of your enablers will include nouns such as *collaboration, infrastructure, applications,* and so on.

5. **Agree on a set of RFID drivers, strategies, and enablers.**

Table 15-1 shows an example of strategies and enablers that are aligned with a driver. Notice how the alignment moves from left to right. The enablers in the table represent some of the core enablers, so you can likely find a few you'll want to include in your own business case. Note how each strategy aligns to the driver, and in parentheses, how each enabler aligns with the strategies.

Table 15-1	An Example of Aligned Drivers, Strategies, and Enablers	
Driver	*Strategies*	*Enablers*
Mandates and compliance: Your company is impacted indirectly through its trading partner relationships. Current key mandates include Wal-Mart and the Department of Defense.	A. Define expectations and requirements with trading partners.	Collaboration with business partners to modify cross-company processes to take advantage of RFID data (All)
	B. Maintain compliance with government regulations.	RFID-tagged items at the unit, case, and pallet level (All)
	C. Develop a transparent and shared view of the supply chain.	Modified applications that integrate and take advantage of RFID data (All)
	D. Make it easier to conduct business.	Transformed business processes that derive benefits from RFID data (C and D)
		RFID infrastructure deployed at appropriate locations throughout the supply chain (C and D)

Step 4. Identify and assess business processes and interfaces

In this step, you develop a high-level model of your company's business processes, which will help you determine how RFID will impact those processes and the associated IS applications.

Although I explain how you complete this step in more detail in Chapter 3, the following steps give you an overview of how to create this model:

1. **Referring to your list of in-scope business processes (refer to the earlier section, "Step 2. Determine scope and assumptions"), construct a high-level, as-is description of those processes.**

    For example, distribution, sales, marketing.

2. **Map the RFID touch points (this is at the business process level, where RFID will have an impact) within the processes.**

3. **Identify the applications associated with the touch points.**

    Think high-level here. Make sure you don't go beyond an initial tentative list of applications.

This step in the business case process is key because it sets the stage for one of the top cost buckets for your business case: the cost of application integration. When you examine the applications associated with the touch points in detail in Step 7, you determine what that cost will be.

Step 5. Identify complementary or competing business initiatives

Here, you establish one of your business case and implementation "radar screens" by identifying what can be leveraged for success and what needs to be closely monitored as competition for the RFID initiative. To accomplish this goal, your team needs to do the following:

1. **Gather information regarding existing initiatives by conducting additional interviews.**

2. **Classify the initiatives as complementary or conflicting with the RFID initiative.**

3. **Develop a table or matrix for complementary initiatives, and another for competing initiatives.**

    Within the complementary initiative matrix, identify actions to leverage and assign core and project team members to either take action or monitor. Within the competing initiative matrix, identify actions to mitigate and assign core and project team members to either take action or monitor.

Step 6. Identify strategic and economic benefits

This step is among the more time-consuming in the process (refer to Figure 15-2). The idea is to summarize the strategic impact and to estimate the potential quantifiable benefits of RFID enablement.

In this step, your team needs to complete the following tasks:

- ✔ **Determine the strategic (nonquantifiable) benefits of RFID employment.**
- ✔ **Quantify the expected process improvements associated with RFID employment.**
- ✔ **Quantify the economic benefits for the expected process improvements.**

You can separate these tasks into two main categories: strategic benefits and quantifiable benefits. In the following sections, I first define what the two types of benefits are, and then I discuss a couple of typical examples and provide practical advice on how to capture each benefit effectively in the business case model.

Determining the strategic benefits

Strategic benefits are the key, nonquantifiable benefits that you test with the core team in order to get agreement on. You can't quantify these, so you need to articulate why they are important to your business.

Make sure you tie strategic benefits to your driver analysis, which you completed already. (See the earlier section, "Step 3. Determine drivers, strategies, and enablers" for details.)

Here are three examples of strategic benefits that you might want to include in your business case:

- ✔ **Transforming business models and methods:** Challenge the team to envision how RFID can take the business to new levels. Ask how making the whole supply chain transparent would enable the company to leverage accurate real-time customer data and thus generate revenue, mitigate risk, and reduce costs. Ask how real-time product sales visibility would enable your sales force to have innovative targeting capabilities, resulting in increased revenue generation and cost reduction.

- ✔ **Reducing counterfeiting:** Because most companies don't want to discuss how counterfeiters have victimized them and at what cost to the brand, reducing counterfeiting is normally a strategic (nonquantifiable) benefit. If you decide to quantify this benefit (using examples of the total cost of counterfeiting incidents) but still present this among the strategic benefits in your business case, there is a logic to follow in communicating the anticounterfeiting benefit:

- The unique RFID tag prevents counterfeiting by providing a way to authenticate the product as genuine.

- The RFID tag raises the bar for the counterfeiter, making activities such as repackaging and relabeling more difficult to accomplish and easier to detect.

- The unique RFID tag enables you to identify reimported product in real time and to intervene.

✔ **Diversion:** Diversion (or rechanneling) is nearly impossible to quantify until you have RFID in place. In your business case, you can state that RFID enables you to identify and act on diverted product and prevents expired returns and other unauthorized product from reentering the supply chain.

The strategic benefits of RFID provide the basis of the business case. Strategic benefits can account for up to two thirds of your overall RFID benefits.

Quantifying process and economic benefits

Quantifiable benefits are the key benefits that you can test and support with calculations. To obtain these calculations, work with both the extended and core teams.

One benefit you may want to quantify is supply-chain efficiencies — although if you're in a regulated industry, such as the pharmaceutical industry, you won't find much here. Here's an overview of how to calculate this benefit, which you can use as an example of how to quantify RFID's timesaving benefits:

1. **Find points in the supply chain at which you believe RFID will make a positive difference.**

   You might look at inventory accuracy in your distribution centers, the loading and unloading of trucks, and the counting associated with the picking and packing process.

2. **Find out how much time is currently spent on these activities.**

3. **For each activity determine a reasonable estimate on timesavings.**

   Use basic math to multiply the annual number of truckloads and orders by the timesavings, and then by a fully loaded rate for the labor.

The benefit to supply-chain efficiencies will likely be modest. Don't spend too much energy on supply-chain efficiency if there already have been plenty of initiatives in this area. Also, people can be very reluctant to agree to quantify benefits that they have had difficulty coming through on in the past.

Another benefit you can quantify is reduced costs on expired returns. For example, RFID can help if your company is reimbursing customers for returns at a greater cost than the original purchase price. In a complicated supply chain with a variety of discounts and contracts in place, RFID tagging at the unit level can sort out when an item was bought and at what effective price. The algorithm that figures these cost variations could be somewhat complicated because of contracts and discounts.

If you challenge yourself, you can establish the logic for a benefit like reduced costs of returns, and then with the help of an extended team member, support that benefit with numbers. Most quantifiable benefits take a good deal of effort to calculate.

As you calculate ROI for quantifiable benefits, keep in mind that prices for RFID tags will eventually drop. Depending on the cost of your items, it's not likely you'll have a positive ROI if you are tagging at the unit level until tag prices drop into the $.08–$.10 range. You can do some "what-if" analysis in your ROI model by dropping the price of tags to find the tag price that gives you a positive return on your investment.

For more on determining quantifiable benefits, see Chapter 16.

Conveying these benefits in the business case

After you collect the necessary information about strategic and economic benefits, you need to convey that information in your business case. Table 15-2 shows how you might present the five benefit areas explained earlier in this section. A level of impact and time to benefit is associated with each benefit:

- **Level of impact:** In Table 15-2, the level of impact takes three factors into account: whether a benefit generates revenue, mitigates risk, or reduces cost. Benefits that meet one factor are low-impact; benefits that meet two or more factors are high-impact.

- **Time frame:** This indicates the time frame in which your business will see benefit. Short term is three to five years; long term is five or more years.

You may also add brief comments to describe how the benefit works, also shown in Table 15-2. Benefits 1 and 2 are quantifiable and will feed directly into the ROI model, as you see later in this chapter in "Step 9. Communicate the business case." For more details on calculating ROI, see Chapter 16.

Table 15-2		Benefits Examples	
Benefit	**Level of Impact**	**Time Frame**	**Comments on Benefits**
1. Supply-chain efficiencies	Low	Short term	Enables only modest efficiency savings within internal supply chain in receiving, picking, packing, and shipping processes.
2. Reduced costs on expired returns	High	Short to long term	Establishes expired returns' actual acquisition cost. Faster, more accurate, targeted recalls enabling definitive compliance with FDA regulations.
3. Transformed business models and methods	High	Long term	Results in revenue generation, risk mitigation, cost reduction.
4. Counterfeiting	High	Short term	Unique RFID tag adds counterfeiting barrier to brands.
5. Diversion	High	Short to long term	Provides capability to identify and act on diverted product. Prevents expired returns from reentering supply chain.

Step 7. Develop investment requirements

Of course, your business case isn't complete without information about what your company needs to invest in order to get RFID up and running. Your team needs to estimate the costs of hardware, software, implementation, integration, training, and support. In the ROI example in Chapter 16, I explain the cost categories you need to address, and I discuss their key characteristics, which will help you estimate accurately.

Step 8. Develop an implementation road map

In this step, you determine the implementation plan in order to understand the timing of the benefits and costs for the business case. The main tasks involved in completing this step are

- ✔ **Developing a pilot concept.** Smaller, controlled projects that sequentially test RFID technology, infrastructure, and collaboration with trading partners.

- ✔ **Develop a rollout concept.** Following piloting, this is the path to production for all products in scope.

- ✔ **Apply benefits and costs according to the timing of the implementation road map.**

Make sure you know when you need to be in compliance with any mandates. If a customer or regulatory body requires you to be RFID-compliant in production at the beginning of 2008, you need to plan around that event. Before 2008, you need to complete, at a minimum, a portfolio of pilot projects and a limited rollout.

Deciding when to implement RFID

When deciding when to implement RFID, consider why you are toying with it in the first place. There are three generally-normally-usually-almost-all-the-time distinct reasons to implement RFID:

- ✔ **Your company has been hit with a mandate to do so.** For example, Wal-Mart insists that all cases be tagged. This creates a concrete deadline, and you need to look back 6–12 months from that deadline and plan accordingly so you that are live on the required date.

- ✔ **You are convinced via an ROI study that your internal business process will benefit in a tangible way from the adoption of RFID, so you want to incorporate the technology as soon as possible to realize that ROI and its strategic benefit.** In this case, the timing can align with your normal business planning cycle and fit in your strategic and budgetary process.

- ✔ **You aren't under a mandate but believe that you will sell more product to a big retailer if you proactively tag your cases.** This instance is similar to the preceding example; however, you'll probably start out by just wetting your feet — to get tags on the minimal amount of products and then evaluate to see whether the potential long-term impact is as good as you might expect.

Although you may believe that the ROI hurdle rate will be achieved in a year or two (or simply feel like hunkering down under your desk and waiting for this RFID phenomenon to pass), there are more reasons to act now than there are to postpone the inevitable. A lot of those reasons center around creating in-house expertise and understanding RFID's impact on infrastructure and business processes. You might need to change everything from packaging to warehouse management systems. Beginning to understand the potential areas for change today will help you plan over the next couple of years when ROI is clearly positive or competitive factors in your industry dictate the use of RFID.

As you consider the timeframe for implementing RFID, you also need to watch closely what your competition is doing. Are they deploying RFID or simply playing with it? Have they gone to their customers with the promise of being RFID-compliant? Does your industry do a lot of business with Wal-Mart, the DoD, Target, or Albertsons?

Timing should be driven by ROI, competitive analysis, client requirements, and potential economies of scale in infrastructure — like doing it when you upgrade your enterprise resource planning (ERP) or warehouse management system (WMS) applications.

Decide whether to implement the entire plan at once

You've heard the expression, "Think globally, act locally." The same applies to an RFID implementation, although in this case, it's "Plan globally, implement locally." Although I emphasize that RFID is a company-level project throughout this book, the implementation is likely to take place in just one or two warehouses or manufacturing facilities. I know you're not a doctor (or haven't played one on TV), but think like a surgeon for a minute. You have a patient who requires treatment in many different areas. As his surgeon, you strategize a treatment plan that will eventually cover all his apparent ills — but you can't operate on them all at once. You might start by operating on the knee, and then on a subsequent day, you might tackle the elbow. In similar fashion, your RFID implementation must be surgically precise.

Because logistics is usually the first part of the company to be confronted with the RFID phenomenon, it is a logical place to pilot the technology. Some companies start with read-only tags where the numbers are pre-programmed. Although this strategy saves a step or two in a slap-and-ship process, it creates headaches later in data management. Read/write tags allow companies to reach back way into the product life cycle and achieve the maximum benefit from the technology. Starting with read/write tags a little at a time is usually your best bet when you go to an actual pilot deployment.

Step 9. Communicate the business case

The last step in the business case is quantifying the overall project and developing a framework for presenting it. You need to produce and include a number of things in your business case:

✓ **A written report of the RFID business case:** Because you build your story as you move through the first eight steps, your team doesn't need to revisit a lot of material at this point.

✓ **A finalized ROI model to support the business case:** The ROI model has been developed along the last three steps of the process (see Chapter 16 for details on actually creating the ROI model). The business case is supported in detail by the information contained in the model.

For each quantifiable benefit and cost category, create a corresponding worksheet within the model.

Figure 15-3 shows an example of a Summary worksheet, which shows the flow of benefits and costs over time (incorporating the road map developed in Step 8). It also shows pie charts depicting the breakout of benefits. Figure 15-3 is just one example of the various effective summary charts and graphs that you can easily create at this point for your business case and presentation. Note that each benefit and cost worksheet feeds the Summary sheet.

Annual Benefits	2005 Low	2005 High	2006 Low	2006 High	2007 Low	2007 High	2008 Low	2008 High
Contract Validation	0	$ -	$ -	$ -	$ -	$ -	$ 4,805,514	$ 6,614,844
Administrative Fees	$ -	$ -	$ -	$ -	$	0	$ 438,578	$ 657,867
Expired Returns	$ -	$ -	$ -	$ -	$ -	$ -	$ 1,991,296	$ 2,212,551
Supply Chain Efficiencies	$ -	$ -	$ -	$ -	$ -	$ -	$ 200,749	$ 288,808
Total Annual Benefits	$ -	$ -	$ -	$ -	$ -	$ -	$ 7,436,137	$ 9,774,070

Annual On-Going Costs	2005 Low	2005 High	2006 Low	2006 High	2007 Low	2007 High	2008 Low	2008 High
Annual Tag Costs	$ 57,500	$ 69,000	$ 185,000	$ 222,000	$ 5,999,680	$ 7,457,975	$ 6,999,816	$ 9,333,088
Annual Maintenance Costs	$ -	$ -	$ 34,730	$ 68,676	$ 152,692	$ 237,230	$ 152,692	$ 237,230
System Maintenance and Mgt	$ -	$ -	$ 1,000	$ 6,000	$ 5,000	$ 30,000	$ 5,000	$ 30,000
Data Storage and Management	$ -	$ -	$ -	$ -	$ 2,917	$ 5,833	$ 2,917	$ 5,833
Total Annual On-Going Costs	$ 57,500	$ 69,000	$ 220,730	$ 296,676	$ 6,160,289	$ 7,731,038	$ 7,160,425	$ 9,606,151

One-Time Costs	2005 Low	2005 High	2006 Low	2006 High	2007 Low	2007 High	2008 Low	2008 High
Application Integration	$ 161,150	$ 193,380	$ 483,450	$ 580,140	$ 690,350	$ 3,539,100	$ 4,777,640	$ 9,047,760
Hardware	$ 182,700	$ 219,240	$ 330,400	$ 396,480	$ 563,000	$ 675,600	$ 2,034,250	$ 2,789,829
Training	$ 30,000	$ 36,000	$ 78,000	$ 93,600	$ 102,000	$ 122,400	$ 262,080	$ 752,640
Software	$ 450,000	$ 540,000	$ -	$ -	$ -	$ -	$ 125,000	$ 450,000
Total One-Time Costs	$ 823,850	$ 988,620	$ 891,850	$ 1,070,220	$ 1,355,350	$ 4,337,100	$ 7,198,970	$ 13,040,229

Figure 15-3: An example of a Summary worksheet in the ROI model.

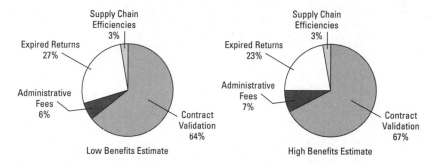

To prepare for your business case presentation, schedule at least two comprehensive sessions with the core team to refine the presentation material. Specifically, you want to develop a project team presentation and an executive-level presentation:

- ✔ **Team-level presentation:** Show a detailed view of the project work, and include ROI model calculations. This presentation is normally used as back-up information if you need to expand on any point during the executive-level presentations.

- ✔ **Executive-level presentation:** Prepare a one page executive summary. Also include recommendations for the RFID implementation and a high-level summary of project work.

For communicating the project, it's always important to have a concise PowerPoint presentation. Bringing this together is pretty simple if you do two things:

- ✔ **Create your presentation graphs and charts within the Excel-based ROI model.** Copy material from the model and paste it into your PowerPoint presentation using the Paste Special and Enhanced Metafile options.

- ✔ **Build your PowerPoint presentation around the nine-step process.** Precede Step 1 with an Executive Summary and Recommendations section, and follow Step 9 with a Summary and Next Steps section.

After the final presentations are locked down, go back and realign the written report as required.

Chapter 16

Fitting RFID into Strategic Plans

Maybe you're already convinced that RFID is the coolest technology on the block, and because all the other kids are getting it, you just gotta have it, too. (If you're the kind of person who justifies another ham radio antenna as an "investment" to your spouse in the event that UFOs land and take out the phone grid, you know what I'm talking about.) Or maybe you're eyeing RFID suspiciously, as you nervously fiddle with your slide rule.

However you regard RFID, the best way to approach your deployment is with good strategic thinking. CEOs, CFOs, and other C-type people are looking for something that will have a positive impact on the organization before they move forward. Coolness just won't cut it. But you can't pretend RFID will ride away on some magic electromagnetic wave, either.

In Chapter 15, I mention that strategic benefits, as well as a thorough return-on-investment (ROI) analysis, are the cornerstones of making a business case for RFID. In this chapter, I focus on the overall strategic impact of the technology. You find out the typical reactions that you might get from employees, partners, and other important stakeholders and discover how to justify the RFID deployment with proven strategic thinking. I walk you through a specific ROI process for budgetary purposes and offer methods for assessing RFID's strategic benefits.

Just in Time to Justify: Overcoming Skepticism with Strategic Thinking

"The bar code can do everything we need to do."

Sound familiar? This sentiment sublimely expresses the feelings of many practitioners who question the benefit of RFID and who believe that complying with their big retail partners has only a cost, not a value-added benefit for their own company. Other skeptics may not have enough information to decide whether RFID is a worthwhile investment. The ROI analysis, which is essential to implementing RFID successfully, may also engender a different form of skepticism that you also need to overcome — pushing for implementation too fast by skipping the ROI study.

When it comes to justifying at the corporate level the insertion and application of RFID technology, you're likely to encounter (or experience) three stages of skeptical behavior:

- **Evasion:** The company might not need to deploy this technology and might be able to wait a few years. Another symptom of evasion is treating RFID like a scourge to pass to the newest executive in the firm because the true benefit is unclear, and no one want to put his or her name on a high-risk project.

- **Denial:** You might think that this technology won't really be that widely adopted, or that Wal-Mart and the Department of Defense (DoD) will back off their mandates. In fact, you might decide that the best course of action is to wait and do nothing. If you are the one with the task of deploying the technology, you might be in denial about the need for a complex ROI study. You might just say, "Hey, we don't need a long, complicated ROI study.

- **Approximation:** If you're trying to figure out the impact on the organization and you know that you are going forward with the technology, it would certainly be easier to just guess at the size of the impact. "It's a really huge problem/opportunity/advantage, man." (I wish I could meet the person who infused the overworked term *huge* into our business vocabulary.) The problem, of course, is that your budget or time allocation might turn out to be entirely wrong if you just approximate. "If we don't start slapping and shipping, we're going to lose market share!"

Of course, recognizing the symptoms of skepticism helps, but to actually overcome it and achieve the company-wide acceptance that your RFID deployment needs, strategic thinking is your best bet. By thinking strategically, you uncover the hard data and information that enables members of your organization to make informed decisions and to communicate the rationale of your RFID deployment effectively. Some people make the mistake of associating an RFID deployment with other IT projects, like Y2K compliance, and just estimate from there. This is a sure way of starting with faulty assumptions.

If you're one of the company's leaders, start thinking and focusing on a sensible business case and ROI analysis that give the senior leadership team a reason to fulfill its destiny in a broad sense, but yet welcome the ability to "surgically" apply the ROI in discrete, smaller areas and processes. If you are the one deploying the technology, welcome the "yardsticks" that a very detailed ROI and strategic plan can offer, so that you know how to measure success and failure.

Calculating ROI — A Tactical Approach to RFID

Return on investment (ROI) is the most talked-about topic in just about every RFID conversation — and it's unique to every company. Doing an ROI study on RFID is a great way to become reacquainted with your company's business processes (BPs). In doing the analysis and performing the calculations, you visit and revisit almost every process in the product life cycle.

As I mention in Chapter 15, completing the business case and calculating ROI go together like the New York Yankees and overspending (did Steinbrenner ever calculate an ROI on A-Rod?). To work through this process, you need to consider some primary areas within your company that will be greatly impacted by an RFID implementation to understand how your business can benefit strategically and economically. These areas range from the impact to your legal department because of governmental compliance issues to changing the equipment and automation in the manufacturing process. Although the impact of the technology is different for each area, the general areas to look at include

- TREAD Act (Transportation Recall Enhancement & Accountability Documentation), Homeland Security, international shipping (compliance)
- Customer requirements (compliance)
- Customer expectations (marketing)
- Customer value (marketing)
- Industrial risk (manufacturing)
- Internal efficiencies (supply chain and logistics)

In the sections that follow, I lead you through a simple, descriptive, easy-to-understand ROI example. You pick up valuable tips on determining how RFID can help make your critical business processes more efficient. Fasten your RFID-tagged seat belt — you won't be able to stop until the ROI is completed!

Cha-ching! Finding ways to save with RFID

This is the section you've been waiting for — your chance to explain in numbers how RFID can save money and increase efficiency. To help you discover potential ways to save money, this section explores a hypothetical company, product, and ROI. If your company is like most of the ones that I've helped comply with high-profile mandates, the first question the CEO or CFO asks is, "What can we get out of it?" When the answer is, "You get to keep doing business with your biggest customer," the next question is usually, "Okay, how can we get value out of it?"

Suppose that you're the owner of Acme Hose & Anvil Co., making hoses and anvils for consumers, auto manufacturers, and pupils of Wile E. Coyote everywhere. The nature of all your products is that they're hard to tag. You can't staple anything to them, bar codes tend to fall off them, and they're hard to store and ship. Labels were the only solution to product identification until RFID came along.

For the ROI example in this chapter, here is some background information on Acme and its hose division:

Annual sales	50,000,000 units
Annual gross revenue	$3,750,000,000
Logistics budget	
Transportation	$300,000,000
Distribution	$100,000,000
Customer returns	$500,000
Average hose sales price	$75/each
Average shipping cost	$5/each

The following list shows some ways in which Acme can reduce its costs through the implementation of RFID:

- Postponement, reduction, and elimination of labels
- Improved shipping accuracy (overage, short, and damage)
- Product age management
- Distribution center, warehousing, and inventory control savings

You can also discover other intangible savings. The following sections discuss these savings areas in detail.

Postponement, reduction, and elimination of labels

Most products require several types of labels for all kinds of reasons: visual product identification, marketing (the pretty labels), processing expectations, and so on. This is often referred to as *human-readable* (the marketing and text stuff) and *machine-readable* (the bar codes) data. For general purposes, the three major sources of product labels are subject to modification or elimination by the use of RFID:

- **Manufacturing labels:** Manufacturing labels can contain stock-keeping units (SKUs), Product Numbers (PNs), in-the-clear text descriptions, legal disclaimers, countries of origin, and other information.

- **Distribution center (DC) and warehouse labels:** DCs can add even more product labels, enabling local less-than-truckload (LTL) transportation processing to occur.

- **Customer (dealer or retailer) labels:** Customers might have their own label requirements that cannot be accommodated by the manufacturer at the time of production because the "build to stock or forecast" manufacturer does not yet know the final destination of the product. Such data includes customer-specific PN and other regional warehousing information. Such labels are occasionally applied by the manufacturer whenever possible.

To determine whether RFID can help to either eliminate or reduce the label workload, you first need to take a look at Acme's costs:

$0.05 cost per label

$0.05 cost of application per label

$0.10 total per label

Acme ships and sells 50,000,000 hoses per year, so the total label cost for one label per hose is $5,000,000. For our example, Acme currently applies three labels to each hose, for a total cost of $15,000,000.

Now along comes the RFID tag. You can implant RFID tags in your hoses and build the appropriate data into them as the hoses travel through the manufacturing and distribution processes. The manufacturer can eliminate at least one label this way and postpone the printing of labels until the optimal time. Through an RFID strategy of label application postponement, Acme can apply one or two labels to the hose instead of three. For this example, assume that you can eliminate one label from the process — that's a $5,000,000 savings and a big process gain.

Michelin has followed a similar process and developed a very cost-effective way to embed RFID tags in rubber tires. This example is a real-world application based on today's technology. Check out Figure 16-1 to see how the RFID tag is built into the tire.

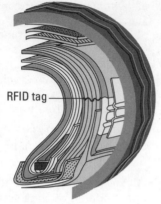

Figure 16-1:
RFID tag
embedded
in a tire
during
manu-
facturing.

RFID tag

Newly developed RFID chips can store a significant amount of information — enough to incorporate manufacturer, third-party logistics provider, and customer data. With RFID tags, it pays for you to postpone printing and applying labels until the last possible minute for the following reasons:

✔ You can add more comprehensive information.

✔ The tags have less chance of being damaged in the manufacturing process.

Print-and-apply solutions are becoming more and more sophisticated as the industry grows. With a print-and-apply RFID label, you can automatically apply the RFID tag, a bar code, in-the-clear text, and graphic logos and images during the manufacturing process — an all-in-one solution. See Chapter 9 for how-to information about applying labels, so you can determine how you might be able to save money in this area and ensure 100-percent accurate reads.

Improved shipping accuracy

Every company wants to improve on logistics. Nothing annoys a customer more than receiving the wrong quantity of merchandise (especially if it's a shortage) or an incorrect product mix. RFID can help your business improve significantly in this area.

Acme Hose currently spends $1,000,000 annually performing physical verification of shipments, either through shipment sampling or a 100-percent manual verification. Add another $250,000 in customer-assessed shipping penalties for errors traceable to Acme. What you (and Acme) need is a sure-fire way to ensure that the right stuff gets on the right truck in the right quantity. Enter RFID.

One of the major challenges in the logistics industry is items that never make it on an outbound truck even though the computer system says that the items are indeed on the truck. This creates a tremendous downstream effect because the inventory arrives at the store (according to the computer system) but isn't really there. The effect is magnified when the store orders more of the item that never arrived when the system says it did. Then the item actually shows up, along with all the extra ordered. Then the store has to cut the price just to get rid of the extra. Whew! What a messs! RFID can fix this nasty bullwhip effect by making sure that what crosses a dock door (pallet, case, or item) is what is associated with a truckload. Here's how:

1. **When a shipment is built and staged, customer-specific shipment information is added to the RFID tag on the item and/or pallet upon which the item is placed.**

2. **Dock door RFID portals read the shipment as the forklift operator loads the pallet onto the trailer.**

   See Chapter 6 for the lowdown on dock door portals.

3. **The middleware or the warehouse management application compares the information that the RFID portal reads with specific and detailed customer order information. If a 100 percent match of quantity and product type is not obtained, the portal sounds the alarm, and the software will determine if there is either a product missing or an extra product on the pallet. If no alarm sounds, the shipper has 100-percent accuracy.**

Customers have certain expectations about shipping accuracy, as well they should. Assume that the following customers and original equipment manufacturers (OEMs) have these standards with respect to shipping accuracy:

Company	Acceptable Error Rates
OEM A	50 parts per million
OEM B	100 parts per million
OEM C	100 parts per million
OEM D	150 parts per million
CUST 1	500 parts per million
CUST 2	600 parts per million

With RFID, Acme can obtain less than 50 parts per million error rate in shipping accuracy — a vast reduction in corrective costs (reshipments, premium transportation charges, customer penalties, lost sales, and so on). A huge boost in customer satisfaction will also inevitably result. By complying to the lowest common denominator in the universe of customers and OEMs, RFID ensures compliance across all trading partners.

So by removing the current $250,000 in penalties mentioned earlier, adding another $750,000 in reduced verification costs to that savings, out of the original $1,250,000 in shipping verification and penalty costs, Acme can conservatively estimate an 80 percent gain, or $1,000,000. Not too shabby, eh? You might make a justification for eliminating all the shipping verification costs and penalties, claiming 100 percent of these savings, but again, for this particular example, just estimate conservatively.

Product age management

Because certain hoses Acme sells have shelf lives, some products cannot exceed customer or manufacturer age constraints before the customer receives the product and the consumer buys it. Manufacturers and distributors in the food, pharmaceuticals, and other consumer product goods industries are sensitive to these requirements, too.

Age lot management is achieved by *clustering* similarly aged lots separately. First-in, first-out (FIFO) techniques are employed on a lot basis to try to ensure that older lots are shipped first. The warehouse management system (WMS) usually considers age in its shipping designation logic. For example, the following product lots are stored in an Acme DC:

Lot Name	Age
Lot A	72 weeks
Lot B	24 weeks
Lot C	40 weeks

The following OEMs have age constraints in receiving products from their suppliers:

Company	Age Constraint
OEM A	52 weeks at time of shipment
OEM B	39 weeks at time of shipment
OEM C	26 weeks at time of shipment
OEM D	40 weeks at time of shipment

After an order from OEM A, the shipper's WMS would logically try to ship Lot A first. However, it can't because OEM A requires the product to be less than 52 weeks old at time of shipment. In this case, the WMS would choose Lot C because Lot C is older than Lot B and satisfies OEM A's age requirement. Additionally OEM C could only receive a shipment from Lot B and should not be sent items from C and A.

Such an age-management system appears easily workable but in fact is fraught with exceptions and errors that creep in when human beings try to comply with what the computer kindly suggests. Here are a few examples of problems that might occur:

- ✔ Products from other lots get mingled with the target lot.
- ✔ The target lot can be split, with part of it conforming and part of it nonconforming.
- ✔ Entire lots can be mis-shipped.

Enter RFID to the rescue. Products' ages are encoded in the RFID tag during the manufacturing process. Just prior to shipment, an RFID read can determine whether the product is permissible for shipment based on an on-the-fly calculation between product birth date and product ship date. Bingo, no more age errors!

In the past, when out-of-age hoses were shipped to customers, Acme had to rectify the situation by reshipping the product and using premium transportation. So 50,000 hoses at twice the normal shipping rate ($10 instead of $5 each) cost the company $500,000.

In addition to saving shipping costs related to out-of-age shipments, RFID can help reduce waste, too. Acme scraps a fair quantity of hoses every year because they are out of age. By using RFID to achieve 100-percent accuracy in the age-management process, Acme can recover revenues to the tune of $3,750,000. This is based on a 0.1 percent scrap rate, or 50,000 of 50,000,000 hoses. Here's the math:

number of hoses scrapped $\times$ average sales price per unit = savings

$50,000 \times \$75.00 = \$3,750,000$

The bottom line is that the 100 percent accuracy in the age-management process saves the company $4,250,000:

$\$3,750,000 + \$500,000 = \$4,250,000$

Distribution center, warehousing, and inventory control savings

Your company can realize gains within the four walls — that is, of a manufacturing or distribution facility — as follows:

- ✔ Eliminate the need to search for lost material.
- ✔ Eliminate time and expense to correct lost material problems through *movement validation* — actually knowing where work-in-process or inventory is at any time in the facility.
- ✔ Save time finding appropriate storage locations through better and faster product ID with RFID.

✔ Maximize product flow through *cross-docking* (shipping directly from receiving without breaking down the pallet), thereby reducing pick-and-stow costs (to pick the items from storage to fill an order and stow them into storage when they come in).

✔ Achieve more efficient pick-and-stow patterns. Better use of task interleaving with RFID-enabled product results in less deadheading in the warehouse and more picks per hour.

✔ Reduce inventory requirements, cycle counting, wall to wall, location surveys, and frequent inventories.

Because savings in this area are dependent on the type and method of DC operation, use the most conservative estimate commonly accepted in the logistics industry. For the Acme example, assume that the adoption of full RFID in DCs and warehouses will bring a 1-percent savings in product handling and storage operations budgets, or $1,000,000.

Intangible areas and gains to consider in your RFID analysis

The RFID savings area examples mentioned earlier pertain to the most obvious and tangible product life cycle processes. The good news is that many other "soft" or intangible areas also benefit from using RFID. Conservatively estimate savings in these areas and add that to your overall company strategy and ROI.

All businesses struggle with weighting various intangible areas when it comes to budget or strategy planning sessions. This is particularly true if a consulting shop veteran or card-carrying MBA leads your strategy and planning events. These folks love hard data and numbers, and if you can't put a metric on it, they don't think it's worth talking about. Yet, as any well-rounded executive knows, intangibles such as client satisfaction, employee morale, adherence to a mission, and audacious goals can separate a good company from a great company. Deploying an RFID network also has its share of intangible strategic benefits:

✔ **Improved customs accuracy:** You can avoid customs penalties by having correct country of origin data codes. Shipping delays for incorrect international shipping documentation can result in significant costs.

Gain estimate: 0.1 percent of transportation costs

✔ **Segregation of rectified product:** In some cases, OEMs and other resellers won't accept *rectified product* (product brought into conformity). RFID can help by toggling a *product rectified* logical field to *true* or *false*. In other words, the RFID tag can let you determine if the product is in conformity or not.

Gain estimate: 0.01 percent of transportation costs

✔ **More accurate plant-level homologation:** OEMs can require that the product they receive originates from only certified supplier plants — such *homologation* is checked at receipt. Internally oriented homologations can also prescribe whether a certain product can be shipped to a particular country, customer, or area. RFID can help by checking against a customer versus plant database at shipment time, ensuring that no errors are made.

Gain estimate: 0.05 percent of transportation costs

✔ **Prevention of questionable product shipments:** If your quality group wants to make some last-minute checks on a product lot, it can ensure that these products do not leave the DCs by coding the database against which the shipment process runs. If a warehouser tries to load a trailer with this product, the RFID portal alarm sounds and prevents the shipment from closing. This prevents the company from expending money and resources to fetch product back while in transit — or worse yet, after having been received by a customer.

Gain estimate: 0.05 percent of transportation costs

✔ **Automation of the customer return process:** Most of what is verified manually can be written into the RFID tag to speed processing.

Gain estimate: 30 percent gain in operational costs associated with returns

✔ **Gray market control:** Products identified with unremoveable and unkill-able RFID tags can be easily identified if they stray from the mainstream product flow into questionable domains. Thieves are less apt to steal and remarket products that can be detected by electronic reads.

Gain estimate: 0.05 percent of gross revenue

✔ **Technology team interest:** This area is one that no dollar figure can be put on, but like the credit card ad says — it's priceless. RFID offers an exciting new technology that the whole technology team can learn and play with. In an environment in which cost cutting is the norm, any new initiative can jump-start the morale of a tired tech team.

Table 16-1 gives a breakdown of Acme's intangible savings areas.

Table 16-1	Summary of Intangible Savings Areas for Acme
Area	*Savings*
Improved customs accuracy	$300,000
Segregation of rectified product	$30,000
More accurate plant-level homologation	$150,000
Prevention of questionable product shipments	$150,000
Automation of the customer return process	$150,000
Gray market control	$1,875,000
Total intangible savings	**$2,655,000**

Sometimes you can put hard numbers on intangibles, and sometimes you can't. When you can't, you need to use an estimating technique everyone can agree on. I'd like to share a technique with you that has worked for me many times in the past. It's called the *halving process*. Here's how it works:

1. **Assemble subject matter experts (SMEs) on the area in question.**

   For example, get your logistics guy, your finance gal, and your IT guru all in the same room.

2. **Describe in general terms what RFID can bring to the table and your best guess as to fuzzy savings estimates in their respective areas.**

3. **From the SMEs' knowledge of specific areas (such as purchasing, legal, personnel, and so on), distill and superimpose their ideas on sales and shipment data.**

4. **Examine the data developed by the SME team and discuss the resultant savings estimates against annual business data.**

   On occasion, when challenged with a soft estimate, I respond by saying, "Don't like that number? Okay, let's cut it in half."

5. **If necessary, halve the intangible estimate and then have another discussion with the SMEs on whether the figure is close to the truth.**

6. **Continue this halving process until all stakeholders can agree on the number.**

An extremely conservative estimate for an intangible gain is always better than none at all.

Tallying up the estimated costs

When you calculate the costs of your RFID deployment, you need to address a handful of cost categories. (Note that this task is Step 8 of the business case methodology I outline in Chapter 15.) In the following list, I discuss each category and its key characteristics, which will help you make more accurate estimates:

  ✔ **Hardware (one-time cost):** If our friends at ACME are setting up 15 interrogation zones across their distribution center, they need to incorporate these capital costs. These costs consist of tag applicators, door readers, forklift readers, hand-held readers, and controllers. Assessing the touch points (points where RFID will impact your business) is essential for accurate hardware costing. (See Chapter 3 for details on determining those touch points and Chapter 15 for details on how you organize your team to complete this task as part of creating a business case.) Some of the variables that you need to determine are what items you'll tag, how many you'll tag, and specific distribution center information (number of doors, forklifts, hand-held devices currently in use for bar code reading). Use the following low and high estimates after you determine where you plan to set up RFID interrogation zones (based on those touch points).

Component	Low Estimate	High Estimate
Tag applicators	$20,000	$100,000
Door readers	$1,500	$8,000
Forklift readers	$8,000	$10,000
Hand-held readers	$5,000	$6,000
Conveyors	$1,000	$7,000
Controllers	$2,000	$3,000

✔ **Application integration (one-time cost):** This is probably the most inaccurate cost category in RFID business cases. The only way to make a relatively accurate estimate is to analyze the list of potential applications that require some degree of integration. (You create this list in Chapter 15.) First, determine how large the integration effort will be by designating each application as either Level 1 or Level 2:

  • **Level 1:** These applications require 200 to 500+ total person days of labor for average Level 1 application integration.

  • **Level 2:** These applications require 50 to 200 total person days of labor for average Level 2 application integration.

Here are some solid numbers to use, based on the labor required. For each Class 1 application, use $300,000 for the low estimate and $725,000 for the high estimate. For each Class 2 application, use $100,000 for the low estimate and $240,000 for the high estimate. These numbers assume a standard internal and external mix of development resources.

✔ **Software (one-time cost):** Middleware is a necessary component for RFID to work properly because it's the engine that makes sense out of the myriad RF reads throughout the supply chain. See Chapter 11 for a more detailed discussion of middleware. For licensing costs, you can safely plan for $200,000 for a low estimate and $450,000 for a high estimate.

✔ **Training (one-time cost):** The best gauge for RFID training for your company is the training you may have already conducted for a warehouse management system implementation. As a general rule, use 6 hours of training for your low estimate, and 10 hours of training for your high estimate. You can then estimate the total cost by using this simple equation:

```
Training hours × Number of distribution center associates
           × Average fully loaded rate
```

✔ **Tags (on-going cost):** Tags are by far your largest RFID system investment, and they are an on-going cost. The first and most important determination that you need to make is how many tags you need over the next five years. In "Step 3. Determine drivers, strategies, and enablers," you discover why you need to tag at either the unit, case, and pallet level or at the case and pallet level. Now you need to do the math. For each product, look at production numbers and sales forecast numbers to understand how many tags will be required.

Because RFID tag costs will come down dramatically, the following table shows how tag costs per unit might look over a five-year period. You can use these cost estimates, along with the numbers of items you need to tag, to determine what the on-going cost will be for five years.

Year	Low Estimate	High Estimate
2005	$.25	$.80
2006	$.14	$.60
2007	$.11	$.45
2008	$.10	$.25
2009	$.04	$.10

✔ **Maintenance of readers and software (on-going cost):** You need to pay annual maintenance costs for the hardware and software, following the first year you procure them. As a rule, maintenance costs 15 to 18 percent of the initial procurement cost (licensing costs for software). Plan on years 2, 3, 4, and 5 for these annual maintenance costs. You'll likely replace the hardware and software in the first five years.

✔ **System maintenance and management (on-going cost):** The RFID system impacts your information system's operating costs. Use $5,000 per month for your low estimate and $30,000 per month for your high estimate.

✔ **Data management (on-going cost):** These costs turn out to be significantly less than you might expect. Assume that each tag is scanned five times between manufacturing and shipping out of the distribution center. Each scan consumes 50 bytes (that is, 16 bytes for the electronic product code [EPC] and 4 bytes for the time stamp, plus an extra 30 bytes for the location, event, and so on). Assuming 100 million tags per year, this yields 500 million records (rows) and 25GB of raw storage per year for event data (not including database overhead). Database overhead is generally 50 percent, which yields 37.5GB per year of necessary disk storage for every 100 million tags read 5 times.

Putting together a costs/benefits analysis

After you account for all the business processes for applicability, both tangible and intangible, you have everything you need to put together an RFID ROI. For the Acme example, you had to generate several assumptions here and there to complete the picture. You now know how RFID will save Acme $13,905,000.

Next, you need to net out the implementation costs and take another look at your ROI position. Figure 16-2 shows an overall three-year costs/benefits picture, using the assumption that individual RFID tag costs will be $0.25 for the first year, $0.15 for Year 2, and $0.10 for Year 3.

ROI as a tool for strategic expansion

As you understand the impact of the technology and get a handle on the specific costs, you need to include critical triggers for expansion in your ROI analysis. Here are some points to consider when scaling up your RFID network:

- **Start with one specific project at a time.** If you adopt a measured approach to RFID deployment, you can assimilate the technology and understand better its impact on your enterprise. After you understand that one project well, you can start to look for other applications in your world.

- **Roll out the applications based on a known ROI or key business metric related to impact.** For example, you can base the timing of two or three of these potential-use cases on ROI. In your ROI, a little analysis will reveal the key drivers in the model (tag price, reader cost, and so on), and you can set triggers for scaling up to those solutions when they yield a positive ROI. So if you know the ROI is break even when, for example, readers are less than $1,000 each or tags are less than $0.10, you can start those projects, or at least put them in the budget, when prices drop to those triggers.

One other critical component determines when and how you scale up your RFID network — your partners. If your partners are also implementing the technology, your benefits can be orders of magnitude higher by closely collaborating with them. The more of your suppliers, customers, and partners you can share real-time data with, the more efficient and real-time your processes can become. And one day you'll get to that nirvana when the network does all the work for you.

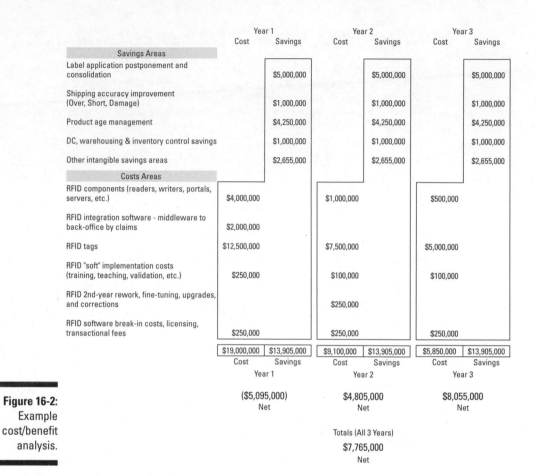

	Year 1		Year 2		Year 3	
	Cost	Savings	Cost	Savings	Cost	Savings
Savings Areas						
Label application postponement and consolidation		$5,000,000		$5,000,000		$5,000,000
Shipping accuracy improvement (Over, Short, Damage)		$1,000,000		$1,000,000		$1,000,000
Product age management		$4,250,000		$4,250,000		$4,250,000
DC, warehousing & inventory control savings		$1,000,000		$1,000,000		$1,000,000
Other intangible savings areas		$2,655,000		$2,655,000		$2,655,000
Costs Areas						
RFID components (readers, writers, portals, servers, etc.)	$4,000,000		$1,000,000		$500,000	
RFID integration software - middleware to back-office by claims	$2,000,000					
RFID tags	$12,500,000		$7,500,000		$5,000,000	
RFID "soft" implementation costs (training, teaching, validation, etc.)	$250,000		$100,000		$100,000	
RFID 2nd-year rework, fine-tuning, upgrades, and corrections			$250,000			
RFID software break-in costs, licensing, transactional fees	$250,000		$250,000		$250,000	
	$19,000,000	$13,905,000	$9,100,000	$13,905,000	$5,850,000	$13,905,000
	Cost	Savings	Cost	Savings	Cost	Savings
	Year 1		Year 2		Year 3	
	($5,095,000) Net		$4,805,000 Net		$8,055,000 Net	

Totals (All 3 Years)
$7,765,000
Net

Figure 16-2:
Example
cost/benefit
analysis.

Tag and You're It: RFID as a Competitive Strategy

Early adopters would say, "How can you *not* justify the adoption of RFID?"
Your business's strategic focus will help you justify the significant investment
required. The well-known Harvard professor Michael Porter formulated a
strategic analysis for industry called the Five Forces model. These five forces
are

- ✔ **Supplier power:** This force includes variables such as supplier concentration, how important volume is to a supplier, and so on.

- ✔ **Threat of substitutes:** Switching costs, buyer's willingness to switch, and the trade-off between performance and price are all examples of this force.

✔ **Degree of rivalry:** Industry concentration and growth, product differentiation, brand identity, fixed costs (money already invested, like for software development or building a data center), and so on dictate the
degree of rivalry.

✔ **Buyer power:** Buyers' ability to bargain, how informed buyers are, the
number of buyers, price sensitivity, product differentiation, substitutes
available, and so on are all part of this force.

✔ **Barriers to entry:** This force comprises the learning curve for proprietary products, access to inputs, government policy, economies of scale,
and more.

Figure 16-3 shows a diagram of Porter's model. (For more information about
Porter's model, go to www.quickmba.com/strategy/porter.shtml.)

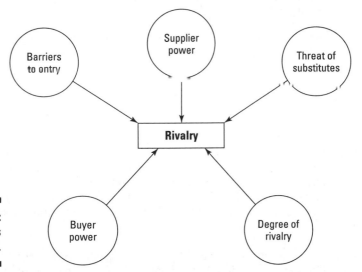

Figure 16-3:
Five Forces
model.

Strategic leaders can apply Porter's model to achieve an acute understanding
of the competitive context in which their companies operate. By so doing,
they can better and more accurately articulate the strategic value of RFID.
For example, an RFID implementation can

✔ **Differentiate your company in a tightly competitive market:** Low levels
of product differentiation within a market sector, whether perceived or
real, are often associated with higher levels of rivalry (competition).
RFID can then be seen as a definite *differentiator* — you have implemented it, and your rivals have not or lag behind. You are the market
sector leader — hooray for you!

✔ **Help a company that is losing market share:** Suppose that your firm is losing market position. According to Porter, this can also cause rivalry intensification. Again, this is a great opportunity for RFID to make the difference — if you adopt ahead of your competitors.

✔ **Increase your company's chances of surviving an industry shakeout:** Industry shakeouts can also warrant special and timely consideration for an RFID implementation. Boston Consulting Group (BCG) founder Bruce Henderson coined the Rule of Three and Four: A stable market will not have more than three significant competitors, and the largest competitor will have no more than four times the market share of the smallest. For more about the Rule of Three and Four, visit

```
www.bcg.com/this_is_bcg/bcg_history/bcg_history_1976.jsp
```

Implications of the rule include the following:

- If there are a larger number of competitors, a shakeout is inevitable.

- Surviving rivals will have to grow faster than the market.

- Eventual losers will have a negative cash flow if they attempt to grow.

- All except the two largest rivals will be losers.

- The definition of what constitutes the "market" is strategically important.

Armed with the preceding information, you can see how important an RFID implementation can be in a *closed* industry, with just a few competitors. Got a product that's difficult to differentiate? Trying to convince consumers that your product is not a commodity? You might want to look at using RFID as your differentiator and come out on top of any shakeout.

One of ODIN technologies' clients was a small supplier to Wal-Mart who volunteered to be part of the first 100 to comply. This move not only got the attention of Wal-Mart's RFID team but also put this company in favored status with the buyers; this firm's business with Wal-Mart has since tripled, and it is achieving a positive ROI on its very simple entry-level RFID program.

Chapter 17

What to Look for When Considering Outsourcing

---

*O*ne of the scary things about RFID and the electronic product code (EPC) is that 60,000 enterprises have to adopt the technology over a three-year period. Those companies have no choice if they want to supply to Wal-Mart, the Department of Defense (DoD), Target, Best Buy, and others. Many more companies have read promising case studies about the technology and want to adopt RFID for some benefit to their bottom line. With all these enterprises adopting RFID at essentially the same time, there are not enough RFID experts in the world today to service even one percent of these firms full-time.

Currently, the RFID industry is so new that only a few providers are able to offer an outsource model. But in the coming years, the industry will evolve with various delivery models, traditional outsourcing firms will get into the RFID space, and specialty providers will focus on outsourcing only RFID.

Because timelines are short and experience is scarce, you might have no choice but to turn your system over to an outsourcing company to install, manage, administer, and monitor. Like giving your daughter away for the rest of her life to some guy you met only a year or two ago, turning your RFID system over to someone else comes with mixed feelings.

This chapter is here to help. First, you need to decide whether outsourcing is truly right for your business and worth the risks associated with handing over the keys to someone else. If you decide that outsourcing is what you need, here I explain what role you should play at the various stages in the project and walk you through the decision-making process for each stage, including

- ✔ Outlining your needs for potential partners
- ✔ Choosing an outsourcing partner
- ✔ Negotiating the service level agreement (SLA) and contract

You also get all the inside secrets from me, a CEO who has been running outsourcing companies for most of his career — just promise not to tell anyone.

Why Outsource Your RFID Network?

Not everyone is a candidate for outsourcing. Step one is to evaluate why you are considering outsourcing your RFID network to someone else. The primary reasons for outsourcing are

- ✔ **Focusing on your core competency:** What you do best likely isn't setting up and running RFID networks.
- ✔ **Lacking resources:** You cannot find and hire dedicated physics and radio frequency experts to keep on staff full-time.
- ✔ **Reaping tax benefits:** By using an outsourcer for "off-balance-sheet" financing, an RFID network becomes an operating expense, which has the potential to limit corporate taxes.
- ✔ **Increasing performance:** Quality outsourcers have the best hardware, software, and systems to optimize an RFID network and usually stay at the forefront of new technology developments.

Identifying and Avoiding the Risks

Outsourcing is not always a bed of roses. Certain risks are involved with allowing an outside firm to manage all or part of your critical infrastructure. The risks can be broken down into four categories:

- ✔ **Technology risks:** RFID is a critical new technology that to many organizations will become as important to their infrastructures as the Internet. Outsourcing the critical component to the wrong partner can put you at risk of technology obsolescence and leave key internal staff members with no understanding of the technology.

Education and regular reviews of milestones and benchmarks are your best defense against failure. You can also limit the technology risk by having one or two members of your team dedicated to keeping up with the technology.

✔ **Core business risks:** If something happens to the outsourcing relationship, the risk to your core business can be considerable. Imagine that your RFID network goes down two years from now, when you are doing most of your business with the Department of Defense, which then refuses to accept any of your products. The time spent finding another provider or fighting for attention when 60,000 other companies are looking for help can be very expensive. This is why it is critical to look at the history of the firm: Make sure that the firm you're considering has been around for a few years and also has a solid balance sheet.

✔ **Business strategy and process risk:** Balancing what is best for the company and what is best for the RFID network is usually not something that the outsourcing provider considers. An example of this type of risk would be if your busiest day of the year is the day after Thanksgiving, but that's the week when your outsourcer chooses to do a system upgrade and test. The responsiveness of a vendor is critical, particularly when it comes time to deliver new products or deploy new systems. Make sure that you put these details in your contract — for instance, not allowing any system upgrades or tests the last week of every month if that's your busy week.

✔ **Operational risks:** With an outsourced relationship, you count on an outside firm to help you meet your deadlines, be fully compliant with a mandate, and not violate any federal regulations. These are tall orders, particularly for people just learning about the technology themselves. Make sure that your outsourcing partner has demonstrated the ability to execute for real-world clients. Look for real-world case studies.

Is Outsourcing Right for You?

I know it would be helpful to give you a couple of quick yes or no answers to the outsourcing question, but like so many information technology (IT) decisions, it's not that easy. What I can give you is a decision process that might help, however. Figure 17-1 shows the decision-making process that your team should consider as you look at outsourcing versus using internal resources.

Although outsourcing is often done because of lack of inhouse *talent,* inhouse *knowledge* to make the right decisions is critical. Chapter 15 introduces key ways to educate your team.

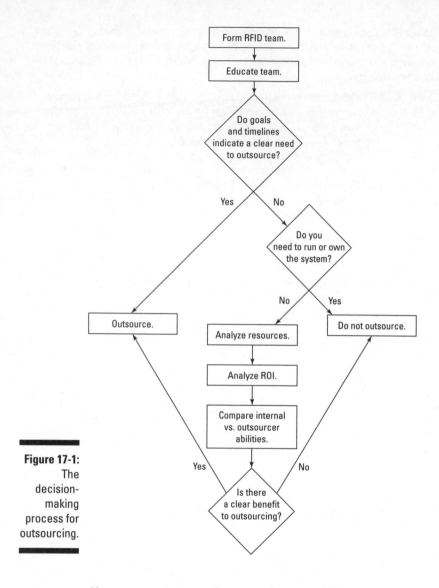

Figure 17-1:
The
decision-
making
process for
outsourcing.

Do your goals and timeline indicate a clear need to outsource?

The core competency and business objectives of your enterprise drive the decision to outsource or not. After getting the team in place, educating your team, and gathering critical data, you can often make a decision in the very early going about whether you need to outsource. This is particularly true if you are under a specific timeline, like a Wal-Mart deadline, and you do not have existing internal resources with any RFID background.

For instance, if, after getting the team together, you find out that IT is having difficulty keeping up with current projects or is in the middle of a new system deployment, it becomes clear that outside expertise is needed. If this is the case, you need to begin looking for a vendor to help you deploy, design, and maintain the system.

The following sections clarify how your RFID team can determine what your goals and timeline are and whether you have the resources to meet those or have a clear need to outsource.

Defining your business goals and timeline

If you are considering outsourcing an RFID network, you need to determine your strategic goals and business requirements — and to do that, you need to know what is involved in the project. That's why education of the team is the key foundation. Sitting down with the RFID team and specifying the goals of the RFID network will help you

✔ Determine the approach you want to take to outsourcing

✔ Decide between using internal resources or an outside firm

In order to figure out your approach and whether you can use internal resources, the team needs to address the following questions:

✔ **What is the goal of the RFID network?** Example goals include slap-and-ship compliance, full system integration, cost savings, and technology advantage. Chapter 15 and 16 help you define what your goals are beyond meeting a mandate. Those two chapters help you build a business case and a return on investment (ROI) analysis. Or you might want to start out with a slap-and-ship strategy and move to a full production system a year or two down the road.

A *production system* implies that RFID readers are located at *all* critical choke points within an enterprise — at the dock doors, on the conveyor belts, at shrink-wrap stations, and so on. It further implies that data is being gathered and integrated into a larger business intelligence or administration system. That system might be warehouse management, enterprise resource planning, inventory control, or other software. The bottom line is that when an RFID network is fully integrated to a production level, it is a key piece of the corporate infrastructure puzzle. If the extent of the RFID implementation is one portal and only a small number of cases are being tagged to meet a mandate, it's not a production system.

✔ **What's the timeline?** You need to know when the network needs to be in place. If you want to move from slap and ship to full production, the team needs to decide when you'll get to a full production system. If the answer is that you won't know until you get there, plan a hypothetical date 24 months out and acknowledge that this date will be flexible. At least having an initial date will help you work back to the design component and help guide your decision-making process.

Can your business meet those goals in time?

Before you can decide to outsource your RFID network, you need to have a good understanding of your business processes in the supply chain, or asset management arena, and also the strengths and weaknesses of your IT and operations groups.

Deciding who will be responsible for the RFID network and what their skills and costs will be is a key driver. Usually, IT or the operations staff is responsible for administration of the RFID network. To make the outsourcing decision, you will need accurate understanding of their capabilities so that you know the level of service required from your outsourcing partner. Some questions your RFID team likely needs to address include

- ✔ Will your business processes have to change if you outsource?

- ✔ Will you have to add access to your facilities or network for the outsourcing partner?

- ✔ What is the internal cost to perform the same operations that your outsourcer will produce? This can include everything from updating firmware on the readers to replacing antennas.

- ✔ How much time will be spent going up a learning curve for your internal team?

- ✔ What other projects could the staff dedicated to RFID be working on?

These are all questions that are likely to clarify whether internal or external expertise should be used.

If, after initial discovery, you cannot make a definite decision either way, more investigation is required. The next two critical decisions are based around owning the network and running the system.

Do you need to run or own the system?

The corporate financing strategy should dictate whether owning the hardware and software associated with an RFID network is critical. This is why I suggest having a bean counter on the team. (See Chapter 15 for the lowdown on putting together the team.) A representative from finance will drive this verdict. Here are some considerations that can help you figure out the answer:

- ✔ **Is it more cost effective to own the system?** Usually, these types of decisions are based on your corporation's weighted average cost of capital (WACC), or how much money costs you over time, versus what it will cost to outsource the same function. Also take into account the difference between leasing the system (an operating expense that is taken out before taxes) and capital expense (which is usually depreciated over three years).

This choice is similar to one that you might make about buying or leasing a car. Leasing the car means relatively little cash out of your pocket to start with, similar monthly payments, but no value at the end of the term; buying it means more money outlay upfront, usually a little more each month, but an owned asset with some value to it at the end of the loan term.

✔ **How will the RFID network fit into your infrastructure?** Deciding whether you need to run the system is a more difficult analysis when considering an RFID network. The difficulty comes in trying to understand where an RFID network will fit into your corporate infrastructure strategy in the long term.

If you are putting in a minimal system to meet one of the big mandates, you might be tempted to decide for simplicity's sake and ease of implementation to just let someone else handle this relatively small project. Two years down the road, however, you might be trying to figure out how to grow the network to include all your dock doors, forklifts, conveyor belts, and other key infrastructure — and be left trying to design an expansive, comprehensive network around the initial deployment that you don't own or run. This will make integration into enterprise resource planning (ERP) or warehouse management systems that much more difficult. Always design with the end in mind.

✔ **Can you afford the risks of switching providers?** Many outsourcing firms focus on business growth by making the switching costs high for their users and "locking them in." If you know that you're going to integrate your RFID network heavily into your IT backbone, you might not want to outsource the data delivery. This would include the middleware or reader interface layer (discussed in detail in Chapter 11). You might, however, want to consider outsourcing the performance monitoring and health of the network, which is one of the most complex components of an RFID network.

✔ **Would outsourcing only monitoring and performance work well for your business?** Outsourcing just the monitoring and performance component has been a successful and proven strategy for IT projects ranging from network connectivity to corporate firewalls. This strategy limits your switching costs and keeps the control of the critical data under your power. It allows you to benefit from the latest technology systems and the best expertise, but limits your risk because you're able to pull the plug on your provider (or vice versa) without shutting down your RFID capabilities. You also have the benefit of capitalizing the hardware.

Always be aware that making switching costs high is a primary goal of most outsourcing firms. *Switching costs* are loosely defined as the pain of migrating from one provider to another provider or back in house. The switching costs can include downtime to the network, resources involved with migration planning, penalties for getting out of long-term contracts, and so on. The better a firm is at making the switching costs high, the more assured it is of not losing customers, even if its service level falls below industry averages.

If you don't need to run or own the system, you've come to the last stage of analysis: measuring three distinct areas of the business to compare outsourcing versus keeping a function inhouse:

- ✔ Resource analysis
- ✔ ROI analysis
- ✔ Performance analysis

The following sections explain how to make these comparisons.

Analyzing your resources

The resource analysis is perhaps the fastest way to determine whether you should outsource. It becomes more complicated with RFID, however, because many enterprises are convinced that RFID is going to be an essential component of the corporate infrastructure and want to maintain control over the system and be empowered to make strategic choices based on internal knowledge.

Before you opt for control, however, you need to answer a few key questions.

Can you meet your deadlines?

The biggest resource challenge to the early adopters is the deadline associated with deploying a system. Your existing IT department might be very anxious to learn about and develop expertise with RFID, but you must balance that desire against a DoD or Wal-Mart mandate that states that you need the system up and running accurately in the immediate future.

Your analysis might clearly show that it would be more cost-effective to use existing staff, but the timeline and other projects underway make that impossible. The physics of RFID are very complex, and many early adopters saw their project timelines double or triple in length after getting into the project, particularly when they had only a trial-and-error process to work from.

If, after investigating internal resources, the timeline to compliance is leaning you toward outsourcing, split your timeline analysis into two resource components — planning and operations:

- ✔ **Planning:** Internal staff needs to be involved in the planning stage at the start of the project to identify scope (what and how much you want to accomplish), define project goals, set up deliverable deadlines, and so on. Because outsourcing means bringing in new resources and team members, the planning time will be more involved than if you were planning an internal project.

For the planning of an RFID network to tag and ship, you should allocate three full months of 8–10 hours per week of involvement from dedicated internal resources. For deployment of a fully integrated RFID network, a good rule is to dedicate 20 man-hours a week for four months for each facility involved in the project.

The time for internal resources will be spent creating the *statement of work* (a detailed description of all the work to be performed), defining and mapping business processes, deciding on location and number of interrogation zones (where you will set up readers and antennas), updating senior management, verifying performance, and being the liaison between corporate and partner entities such as Wal-Mart.

✔ **Operations:** The operations role required of your internal resources will not actually be installing and setting up the systems because you're relying on someone else for that. However, internal resources are still needed for projects like wiring power, running cable, and planning the integration with your IT systems. The lion's share of the effort will be spent getting the data from the RFID readers and converting it into something useful for your business applications to use. This requires your internal staff to understand how data is exported from the middleware (also called the EPC information services or Savant, the piece that filters and smoothes all the reader data) layer and if that middleware is included in the outsourcing arrangement.

Can internal staff handle the demands of RFID?

If you are one of the proactive thinkers investigating and deploying an RFID network before you are required to, you have the luxury of beefing up internal resources to take ownership of the project. This has a number of advantages. First, it keeps control and understanding on your side. This is critical even if you decide to outsource so that you understand what is involved in deployment and management. Secondly, it gives your internal staff an exciting new technology to play with, a motivator that you might not want to overlook. Giving employees new technology to work with can often be more powerful than salary or stock options when it comes to job satisfaction. Having the chance to work with RFID can breathe new life into a technology organization that might have spent the last several years cutting costs, people, and projects in order to weather a bad economy.

The most important issue that you need to consider is the IT staff's current utilization. As motivating as a new technology can be, if it's another project placed on top of someone who is already putting in 60-hour weeks, you could have a recipe for disaster. Find out upfront how much flexibility your IT has to take on new projects. For example, are other projects ending, or is there allocation for new staffing? The answers will also help determine whether your internal staff can handle the stress and excitement of a major new infrastructure project.

If you decide to use internal staff to get up to speed on RFID, you should look for folks with electrical engineering, physics, or computer science backgrounds as a solid foundation for learning radio-frequency basics.

Money, money, money: Comparing outsourcing and internal costs

The second decision-making criterion is the ROI or cost-benefit analysis. In Chapter 15, I talk about a very pragmatic approach to determining the return on investment on an RFID project. The outsourcing ROI analysis is slightly different in that it assumes that you *will* deploy the RFID network. Now what you are comparing is internal costs versus outsourcing costs. To figure out the ROI analysis for the outsourcing model, you need to determine whether you are going to own or lease the hardware, and how much of the infrastructure will be operating expense versus capital expense. This is where the CFO on your team comes in handy. Taking the basic business case and adapting it for outsourcing is a key project for the finance team.

Keep in mind, however, that although the ROI might not be greater for an outsourcing option, the risks might be lower, the timeline faster, or the impact on other projects lessened. That is why the ROI analysis is only one of three parameters (along with resources and performance) to consider at this stage in the analysis.

To compare internal ROI and outsourced ROI, the statement of work and project plan have to be available and detailed enough for your finance team to see all the costs as well as how the costs differ from outsourcing versus inhouse endeavors. In the analysis, look at four key components:

- ✔ **Total project cost:** This includes deployment time, certification, hardware, software, business process reengineering, and so on. When considering internal resources, you need to include the significant time and cost of training, education, and if necessary, recruitment of new employees with specific expertise.

- ✔ **Annual maintenance costs:** This is the most difficult to assess accurately because RFID is such a new technology. However, if you outsource to a reputable vendor, you will know exactly what the annual costs will be.

For internal purposes, you can choose a multiple of your typical IT maintenance costs. For example, if your TCP/IP network has an annual maintenance cost of 25 percent of total project costs, you might want to use a multiple of 1.5 or 2.0 times that maintenance cost for the first couple of years of an RFID network to account for new technology and a learning curve.

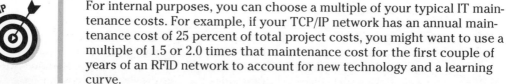

✔ **Risk-based costs:** Considering what happens when things go bad is one of the more esoteric components of an outsourcing analysis but one that is important to try to quantify. An example of a risk-based cost is the cost of keeping the system internal to the organization. Another example is purchasing the hardware, software, and tools, and then having a protocol or standards change that renders all your technology obsolete. See Chapter 2 for details on protocols and standards.

A risk-based cost that is easier to quantify is failure to comply with mandates because of poor RFID engineering. If product going to big retailers does not have a tag on it when the systems go full production, the suppliers will be assessed a penalty for each noncompliance case. It's worse with the DoD; those guys will just flatly refuse to accept a shipment that is not readable from an EPC-compliant RFID system. These risk-based costs can be put on the shoulders of an outsourcing partner.

✔ **Management costs:** No matter who does the project, internal resources play a critical role in program and project management. If it's outsourced, you will require 45–60 percent fewer resources than if the entire project were deployed internally.

The four cost categories should be detailed over a five-year period, which accounts for your expansion as the technology matures and the time value of money (your corporate WACC that I mention earlier in this chapter).

Performance anxiety: Can you build a network that works?

The third and final criterion for deciding on an outsourcing option is whether someone can do it better than you can. Although RFID is not a new technology, today's RFID and EPC systems are new applications of the technology.

The emergent nature of the RFID market means that not a lot of experienced people are available for companies to hire internally. Many companies that were lucky enough to have internal RFID experience have seen their experts leave to start RFID-specific consultancies because their knowledge is so valuable in today's marketplace. The lack of experience, coupled with the early stage of products, has created a distinct performance advantage for people with extensive RFID experience, specifically as it relates to today's RFID readers, antennas, and tags.

The handful of RFID-specific companies that have been around since the early 2000s clearly have a significant advantage over firms started in 2004 and 2005. The older companies have experience with the hardware and with actual deployments. The top companies have leveraged that experience to create a scientific methodology as opposed to trial and error. With these advantages, such companies can easily cut the deployment time in half and greatly increase the performance of your network.

It is important to have an advantage in time and performance if you are under an early mandate and have a short timeline to adopt RFID. If you're not required to have something up and running but are investigating RFID because you feel it will be a strategic advantage or necessity in the near future, you might not need the expertise of an experienced RFID expert because working with RFID will only get easier.

The bottom line of the performance issue is that if time is a factor, you will not be able to get up to speed quickly enough internally. And if you are trying to get away with a minimalist slap-and-ship system, when you do evolve into a pervasive RFID network, your design and ongoing maintenance costs will be significantly higher because you did not start with the end goal in mind.

Finding the Perfect Match

If you're planning to outsource your RFID network, two factors will make or break your future:

- ✔ The partner you choose
- ✔ Your internal level of knowledge and how well you keep up with the technology

Pay special attention to note the word *partner:* An outsourcing relationship should go beyond a typical vendor/client affiliation. The RFID network, when fully integrated into your business process, has the potential to completely grind shipping and production to a halt if something goes awry. If you are relying on an outsourcing partner to make sure that your RFID network is healthy, trusting the right one is critical.

Unfortunately, matchmaker.com doesn't have a section for RFID mates, but the following sections can help.

Figuring out the RFP process

You find a vendor by creating a request for proposal (RFP). This can be an arduous process, but it's worth the trouble in order to develop the right relationship with the right partner. The following steps outline how the RFP process works:

1. **Work with internal staff and possibly a consultant to gather the information about your business and your needs so that you can put together a detailed RFP.**

   The data-gathering and discovery process that your RFID team has gone through will help you glean a much better understanding of your corporate strengths and weaknesses relating to preparedness for an RFID network.

   If you've already worked through the questions in the previous section, "Is Outsourcing Right for You?," you've done much of the work already. If not, see that section for details.

   Map out the various stages of the business process on large sheets of paper where you can identify the locations of RFID interrogation zones and where business processes might be changed. (Chapter 6 talks about locations where you might put an RFID reader or interrogation zone.) These maps are very helpful as you try to understand the long-term implications of the system. This process mapping can be seen in Figure 17-2. A process map is a great visual foundation for an RFP.

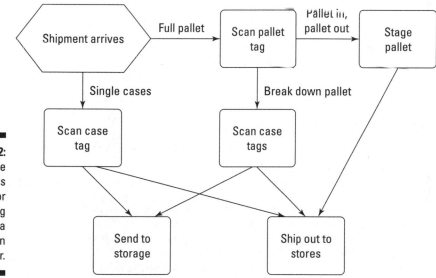

Figure 17-2: A simple process map for receiving into a distribution center.

2. **Create a detailed RFP.**

   When you create the RFP, you put all the details that you've collected about your needs and how much or how little you expect the partner to do. By spelling out all these details, you're much more likely to find a partner that can meet those needs — and thus have a successful relationship.

3. **Send the RFP to several different kinds of service providers.**

   I offer tips about this later in this chapter. The Web sites in Chapter 19 can also help you find reputable companies.

4. **Evaluate the proposals that vendors send you.**

In the next few sections, I explain what happens each step of the way.

Spelling out your needs in an RFP

The RFP should focus on both long- and short-term requirements so that your partner can design a system from the start that will scale as your RFID requirements grow. Table 17-1 offers a sketch of the components you need to include, in the order you likely want to include them.

Table 17-1	Putting Together an RFP
Component	*What It's For*
Organizational Overview	This helps vendors understand who will be involved and where the focus is in the organization.
Required Deliverables	This is the meat of your RFP, where you outline technology requirements, monitoring and performance management, support, and deployment locations.
Required Proposal Format	Making sure you hold vendors to the same format makes an apples-to-apples comparison easier on you, and also helps you determine which vendors listen to your requests.
Request for References	A critical component to see who is full of marketing fluff and who has actually done the work for *real paying* clients. Check these thoroughly.
Mutual NDA (nondisclosure agreement)	You want legal protection as you share details about your business and ask potential partners to share details about theirs.
Submission deadline	Set a reasonable deadline for submissions and hold to it. Then give accurate expectations to the vendors, when there will be a short list notification, and when finalists will come in for oral presentations.

In the following sections, I explain the more complex components of the RFP in more detail.

Technology requirements

If the cost is the potatoes of an RFP, the technology is the meat. Technology drives many of the other considerations, from selection to timing. Because many of the specific technology components are covered in other places in this book, I'll give you just a few reminders for the various parts that make up the technology section of an RFP:

- **Product testing:** What are the methodology and processes used to test for optimal tags and tag placement on your products being tagged? There should be a demonstrable use of scientific method and not just trial and error. Ask to see examples of past reports as an appendix. Detail your total number of products and how many you plan on tagging or testing. I talk about testing the products for the right tags and placement in Chapter 9.

- **Site assessment(s):** Make sure that a Full Faraday Cycle Analysis (see Chapter 7 for details) is part of the evaluation. Ask to see examples of past site assessments and see whether there is information that the deployment team can use to help configure and tune the various hardware. An RF path loss contour map (see Chapter 7) is usually used as a design foundation.

- **Hardware selection:** Because there is no silver bullet for RFID, ask whether the firms have experience with multiple hardware vendors so that they can provide the best hardware for a specific location within your RFID network. Ask what relationships the outsourcer has with various hardware manufacturers; make sure that they have four or five companies to choose from and also that the outsourcer has multiple engineers certified through the various hardware products.

- **Network design:** This is critical for long-term success and for minimizing ongoing maintenance costs. The RFID network should be designed with a three-year time horizon in mind. In other words, what do you envision the network looking like in three years? Today, it might have only one or two readers set up, but you need to look to the future to figure out the long-term vision. Key consideration should be given to understanding communication needs such as Wi-Fi, Ethernet, RS-485, and others.

- **Hardware testing and certification:** Because RFID readers and antenna production is in limited quantities today, quality control isn't at the level of a mature industry. Therefore, ask the outsourcer to test every reader in a controlled environment to ensure integrity of the RF signal compliance with the rules of the Federal Communications Commission (FCC). The vendor should be willing to sign off on accuracy of testing to protect you from any liability associated with FCC compliance. See Chapter 8 for details on testing.

✓ **Hardware installation and on-site testing:** Installing the readers is tougher than most people think. The installation has to be flexible enough to allow design and configuration flexibility yet durable enough to withstand an industrial environment. Does the company make or have access to well-designed RFID racks and enclosures? Ask to see photographs of recent installations.

✓ **Software requirements:** Chapter 11 details the middleware aspect of an RFID system, but asking for successful deployment references is critical. I once worked with a leading middleware company for six months with its "mature" product and still couldn't get it to work on the network that our client designed. Demos and the real world are as different as night and day. If a single outsource partner is going to handle physics, network design, and middleware, that partner is likely to use someone else's middleware. In response to the RFP, the partner might use the middleware company's installations as its own proven references. Find out what the partner has done directly as well as how many people in the organization are trained on the middleware that the partner will likely use.

✓ **Forward compatibility:** This is something many people focus on but few address in an RFP. A key benefit of outsourcing is that someone else shoulders the risk of technology obsolescence. Your RFP must clarify that if the protocols, standards, or technology changes, your partner will be responsible for making those changes to keep you up-to-date. This will guarantee that your potential partner isn't buying older hardware, which might not be forward compatible.

✓ **Security issues:** Depending on your product and environment, security can be a vital part of your decision. Consider the two forms of security:

- *Technical security* takes into consideration how things are encrypted and password-protected.

- *Operation security* deals with the people. Look for a company that does background checks on all employees and ask for the name of the firm that performs the investigations if security is a genuine concern.

The best proxy for in-depth due diligence is to find out whether the company has a U.S. government security clearance.

The gold seal of approval from a security perspective is if the company has a U.S. Federal Facilities Clearance. If so, the company should have a dedicated facilities security officer (FSO). Ask for the name of that person and call the Defense Industrial Security Clearance Office (DISCO) at 1-888-282-7682 to verify for yourself whether that person is in fact the FSO.

Monitoring and performance management

Monitoring the health and welfare of your RFID network becomes increasingly important as the network expands and its importance to your business process and decision-making increases. Outsourcing can give you better access to the latest and greatest monitoring software. The performance monitoring is a critical part of the service level agreement (SLA, which I discuss later in this chapter). Make sure that when you put together your RFP, you spell out the expectations of the SLA.

The level of monitoring and performance you require depends on the importance of your RFID system. If many of your outbound shipments require RFID tags, make sure that you have a system that is up and running during your normal production hours.

Up and running is a fuzzy phrase that can be defined in many ways. Chapter 14 details the best methods for monitoring a system. However, be keenly aware of measuring performance instead of just measuring aliveness. If your RFID network is critical to your business and you face penalties or nonpayment for not having a valid RFID tag on each case, demand an SLA that supports at least 99 percent availability and notification within 15 minutes of any issue. The RFP must clarify that any vendors will be required to provide a 99-percent uptime level of service and will be expected to sign an SLA to enforce penalties for nonperformance.

Technical support

Before you decide to let your precious RFID network into the hands of another, you need to be assured that it will be taken good care of when something goes wrong. The support section of an RFP should be detailed and comprehensive.

From a support perspective, most IT projects benefit from having one throat to choke — or as I like to say, one back to pat. This means that you should be able to call a dedicated toll-free number if an issue arises. This is especially important if you are not outsourcing performance monitoring and management, which is by definition a *reactive approach* — that is, you have to wait until something breaks and then decide how to react to it.

Support provided by the vendor should include

- ✔ **Access to a toll-free (800) customer support number** that is available during your hours of operation (a single 8-hour shift or 24/7), regardless of your or the outsourcing partner's location. For outsourcing vendors located outside the United States, make sure that support lines include support in the language of your operations team.

- ✔ **E-mail access** to customer support.

- ✔ **Scheduled maintenance windows** that are outside your critical business hours. Ask vendors to conduct system patches, upgrades, and fixes during the scheduled window — unless there are security or system risks, defined as Level 1 problems (see the upcoming Table 17-2).

- ✔ **Support for all application software,** including diagnostic services and software patches.

- ✔ **Support for system hardware,** including diagnostic services and hardware vendor contact assistance.

- ✔ **Support for any third-party software,** including diagnostic services, software vendor contact assistance, and software patches.

Other support features that you might want your outsourcing partner to provide include

- ✔ **The ability to catalog a knowledge base of resolutions for historical reference:** This can be a great resource that you can keep on hand rather than having to wait for a response from support.

- ✔ **Monthly call reports:** Your outsourcing partner might send these out electronically or make them available over the Web. The reports give you the opportunity to review the call history for proactive management of issues.

- ✔ **A list of several groups and their associated roles:** Typically, the four groups available in customer support are system engineers, database administrators, performance analysts, and project engineers. Within these groups, individuals are sometimes divided into various tiers depending on skill level (tier 1 being a trained call center person, up to a tier 3 individual who might actually work on the systems in deployment or development).

- ✔ **A Web-based portal:** Some of the larger outsourcing firms make resources available via a Web-based portal so that you can get to information about your RFID network (such as trouble tickets and performance metrics) anytime you want.

Even if a Web portal isn't available, find out what the outsourcing company's trouble-ticket or issue-tracking system looks like. It doesn't have to be an expensive brand-name system like Siebel, but it does have to easily track the status of a specific case by trouble ticket number, quickly add information to an existing case, and automatically escalate a problem if it's not solved in a certain amount of time.

Deployment location (s)

The RFP needs to include all locations contemplated for the initial and long-term deployment. A company will price the first piece of business much more favorably if it knows that it will deploy several more of the exact same systems in the near future.

Spelling out all your locations not only gives you better pricing, but also helps your prospective vendors understand potential support issues associated with varying time zones, different languages, and likely communication constraints. This also helps them to accurately gauge travel expenses and response time to remediate support issues.

Experience and references

Experience and reference customers are the key to seeing how successful the outsource partner has been in real-world environments. This is the make-or-break section of your RFP. Be sure to ask a potential outsourcing partner for references. You also want the company to provide details about its RFID experience.

Many an outsourcing provider will claim that all its clients are covered by nondisclosure agreements (NDAs). Although this is certainly true in many instances, at least one or two customers should be able to talk about the vendor's ability to provide a service. Beware of a company that doesn't want to provide references.

To assess the quality of an outsourcing partner, ask for details about the senior leaders in the organization as well as the people who will be working directly on your project. With RFID getting so much press, a lot of opportunistic people have started RFID firms. To get a good idea of whether an outsourcing vendor is worth your time and money, I recommend asking the following questions:

- **How long have key personnel been working with RFID?** The best firms have key personnel who have been in the business for decades, not months. Look for chief scientists, VPs of engineering, and similar-level positions to be held by people who have spent years with different radio-frequency systems.

- **Where did they get that experience?** Many great practitioners have come out of various government programs like NASA, think tanks like the Johns Hopkins University Applied Physics Laboratory or the MIT Auto-ID Center, or specific educational programs like the Case Western Physics Entrepreneurship Program. Also, several companies have a history in radio frequency — Texas Instruments, Motorola, Intermec, and others — which are good breeding grounds for RF experiences.

- **What does the company do besides make money?** This is the touchy-feely stuff that really does matter. Good people make good companies. Asking this question in the RFP can give you some insight into how Machiavellian a partner might be when dealing with you. Look for firms that are involved with standards bodies like the International Organization for Standardization (ISO), the Internet Engineering Task Force (IETF), EPCglobal working groups, or other such bodies. Also, companies that participate in associations and industry groups are likely to be better connected and more on the pulse of the technology. Lastly, if a company believes in taking care of all the stakeholders in its area, whether by supporting local charities or by giving employees time off to volunteer, it usually means that senior management values things other than how high they can drive their stock price.

Selecting potential outsourcing partners

After you craft and then review your RFP to make sure it's ready to send out, you have to decide where to send it. The Web sites in Chapter 19 can help you find specific companies. I recommend sending it to at least four or five vendors; look at the big global firms as well as the small specialty shops. Here, I offer more general advice:

- **Know what to expect from small companies versus big companies.** Keep in mind that large companies need large contracts to pay the large bills. IBM and Accenture have much greater overhead and administration costs than ODIN technologies or OATSystems. Therefore, the larger companies are more likely to provide a higher level of service to the clients who have contracted for multimillion-dollar deployments. A $200,000 contract will not have any effect on the bottom line of a company like IBM, so expect its account managers to try and increase the margin on smaller customers by charging for service calls, increasing maintenance contracts, and other nickel-and-dime items. Another way these big companies can justify taking on a small customer is to develop a product for your deployment that they can reuse with other clients.

  The small company/big company decision is largely a function of your need to get personalized service and specific experience balanced with your desire to have a company that is well recognized and financially significant. Either way, reference customers hold the key to unlocking the true service capability of the vendor.

- **Find out who is backing the company.** There are several types of companies in the RFID space today. Examining who makes up the company and the board of directors is a good way to get insight into the goals of the company. Look for firms that are made up of passionate RFID experts — people who might have left large, stable companies to start their own companies.

  If you see a lot of venture capitalists and if the company was started post-2003, it is likely comprised of opportunistic types looking for a profitable and fast exit. Such a company might not exist in a year or two. The situation is similar to the many data-hosting companies started in the late 1990s: These companies provided great customer service and responsiveness, but as soon as they were sold to large conglomerates, the service and innovation immediately degraded.

- **Consider where RFID fits in terms of strategic importance and what else the company might be doing in addition to RFID.** Nobody ever got fired for hiring Big Blue, but the RFID team lives somewhere down at the bottom of the pecking order, after data hosting, blade servers, storage systems, strategic consulting . . . well, you get the idea.

Evaluating responses to your RFP

After you send out your RFP, companies that are interested in providing you with services will respond with varying levels of detail by providing you a proposal and cost breakdown. When you send out your RFP, remember to put a deadline on the responses and then hold to that deadline. After all, if a company can't keep to a reasonable deadline for an RFP response, how will it get your project finished on time?

Here are some tips that can help you separate the good responses from the bad:

- **No references, no deal.** If you get an RFP response back that doesn't list reference customers and contact names (assuming that you asked for them), don't bother looking at the rest of the proposal. Eliminate the firm from your consideration unless you want someone learning RFID on your dime.

- **Come up with the cash.** Always remember that companies have to be self-sustaining. It is perfectly acceptable for a company to make a reasonable profit and be able to pay their employees a fair wage.

- **Scrutinize lowball bids.** If you get a proposal back that is significantly lower than the rest, the company is either exhibiting bad business and might not be around in the near future, or it plans on sticking precisely to the letter of the RFP and charging you extra for every minor change in scope. Most IT professionals have learned the hard way that when Big Company offers something for free, that same Big Company makes it up (and then some) in other services. Watch out for this!

Sealing the Deal with an SLA

After you find an outsourcing partner you want to work with, you're ready to move on to the service level agreement (SLA), which clarifies to both parties exactly what services will be provided and what penalties will be assessed for noncompliance.

An outsourcing partner can be your knight in shining armor if you need to meet a compliance deadline. But making sure that you take the time and effort to lift up the visor and see what's under the armor is your key to being happy with the relationship. Creating a sound RFP will evolve into the foundation for an accurate SLA.

The SLA is the legal agreement that makes your partner sign on to all the things promised in the response back from the RFP. In other words, if the outsource partner claims 24/7 support in its RFP response, the SLA needs to spell out the penalties the partner will pay if you cannot actually get service 24/7.

To get to an effective SLA, follow these steps:

1. **List any metrics included in the RFP.**

   Metrics are simply items in your RFP that are measurable and concrete, including time to respond to trouble tickets, definition of problem severity, e-mail or phone call acknowledgement of problems, and so forth.

2. **Rank the order of importance of those metrics from most important to least important.**

3. **Work with the finance and RFID team to figure out the effect of not achieving the desired levels on the various metrics.**

   In other words, if the RFID network goes down and no cases are being tagged and read, would you stop production or shipping? Would you be charged a penalty from your customer? If the answer is yes, what does it cost for every minute the production line is shut down, or how much of a penalty will your customer charge you?

4. **Take the costs of nonperformance and put them into contractual language so that your provider absorbs those costs if the SLAs are not met.**

5. **Negotiate the SLA with the outsourcing provider.**

   Many outsourcing providers have a standard SLA. You can ask them to send that back in their response to your RFP so you could see what they are really willing to commit to versus what they say in their marketing and sales pitches.

Drafting the initial SLA

The service level agreement sets the objectives and understanding of the partnership. When you are outsourcing a key infrastructure component, it really is a partnership — and like any partnership, you want to know what to expect during the good times as well as the bad. After all, Shrek is about the only one I know who was happy that the beautiful princess turned into an ogre at midnight.

The implementation/deployment phase of the contract (to meet a deadline) uses different metrics than the on-going performance and monitoring (after you're up and running). I suggest splitting the two phases into two separate SLAs with specific penalties for failure.

Here's what your SLA needs to accomplish:

✔ **Set service-level benchmarks.** In order to obtain the best service possible, keep in mind that you aren't buying capacity or technology but rather a service level. This means that you need a good understanding of performance measures and metrics.

✔ **Ensure good service with penalties.** The best SLAs have significant penalties for noncompliance to make sure that your outsourcing partner takes the commitment to heart. SLAs that have no teeth offer very high service levels and response time guarantees, yet no penalty for noncompliance.

Here are some tips for setting the right service levels:

• **Use the same metrics for performance measures that you used in the RFP's technical section (read accuracy and response time, for example).**

• **A good performance measure is a service that's better than you can do yourself or are getting now.** For example, one of your performance measures might be getting a 95-percent read success on all outbound cases. For not meeting that performance level, you might give them ten business days to fix it; and if they don't fix it, you may ask for a $5,000 penalty for each failure. This also assumes that you are going to pay your outsourcer for a full maintenance contract to support this kind of service level.

• **Make sure that the vendor's performance is aligned with your needs as an organization.** Always keep the big picture in mind when working through the details of an SLA. If your business goal is better performance or technical advantage, your SLA should also include incentives to encourage the vendor to meet those goals.

✔ **Ensure that your outsourcing provider creates and maintains back-up files of your reader configuration and version.** The configuration of each reader is a valuable piece of information that your service provider should store in a secure database that includes weekly tape back-ups in case the server crashes.

If the SLA correctly scopes out the expectations of services, you should not see any creep in the project costs.

When you articulate your support requirements, it is reasonable to expect that the outsourcing company will offer different SLAs based on the severity of the problem. Normally, problems are broken into three or four levels of severity, which are outlined in Table 17-2. Note that you receive notification only if performance monitoring is outsourced.

Table 17-2   Levels of Severity for RFID Network Problems

Level description	When should support notify you?	What's the workaround or solution?	Examples
Level 1 is the most severe type of problem: a critical failure in operational activity.	Within a very short timeframe — under 15 minutes in most cases	Because these are show-stoppers, no workaround is available.	The failure of the main RFID system hardware or operating system, RFID network failure to report data to applications, database failure, or major application subsystem failures.
Level 2 is not as severe as the impact of a Level 1 failure but still is something that requires immediate attention, if not immediate resolution.	In less than 30 minutes; resolution should come during the same business day	A Level 2 issue usually has a short-term workaround available, such as rebooting the reader and reapplying the proper configuration, or shutting down and restarting the database.	The degradation of the performance of the main system, usually measured by the monitoring system. This can be one of the scary times in which RFID hardware still pings that it is up and online, but behaves outside the normal pattern, indicating an inoperative or seriously degraded application.
Level 3 is a problem of mild impact, if any, to the overall system. The problem limits functionality or usefulness of an application but isn't critical to the system operation.	During a scheduled maintenance window	A workaround is readily available and can be applied or used with little or no operational impact.	The failure of a given component of a subsystem component. This can include an enunciator light, which notifies of successful reads, or a broken antenna housing.
Level 4 doesn't affect system operation or data integrity. It is a minimal problem arising from a misleading or unsatisfactory component or feature.	During or after the scheduled maintenance window	These are usually addressed in scheduled maintenance periods. Deferred maintenance of the Level 4 problem is acceptable.	An improper field label, a misleading or inappropriate error message, or the failure of a noncritical, display-only form.

Many firms, particularly those with large consulting arms, purposefully make these components vague so that they can either sell additional services and consulting or use the SLA as a lever to upgrade services and increase their revenues. This happens fairly often with low bidders in the RFP process, particularly if they are big consulting or technology companies.

Negotiating an SLA with a vendor

After your team has sorted out the various RFP responses and put some metrics into a format for measuring performance and compliance, you are ready to start the negotiating process.

First, you need to assemble your negotiating team. If your company has a procurement expert, this is the time to get him or her involved. If not, legal counsel is usually helpful — but plan on taking a week or two for legal review. The members of your RFID team are also critical at the negotiating table because they understand the core requirements from a technical, operations, and business process perspective.

Smart vendors find as many contact points within an organization as possible, always working back channels for more information or to determine who is making decisions. You can avoid this potentially confusing covert information gathering by asking your team members to direct all inquiries from vendors to the main point of contact in your firm. Also keep in mind that you need to share as much information in the RFP as possible so that the potential vendors don't have to covertly gather information. I never understand why people are reluctant to share information when looking for a solution. After all, you wouldn't ask your doctor to guess what your symptoms are. A technology partner needs to make the right diagnosis as well, so share all you can — just make sure you're protected under an NDA.

As you negotiate the contract and SLA, prioritize what the must-have components are for you and try to separate them from the nice-to-haves. Then put yourself in the vendor's shoes. You find out plenty about your vendor from the RFP process, so what do you think costs the most amount of time, money, and effort? What things are relatively easy for the vendor to do? Don't be afraid to ask these questions directly. The answers are your levers in the negotiation phases. Understanding what you need and what your vendor can provide relatively easily will help you both achieve a win-win outcome. An SLA negotiation should be the beginning of a great partnership, not the territorial battle some lawyers like to make it. It is better to be cooperative than adversarial.

A collaborative negotiating stance will ensure that the SLA is achievable and enforceable. Sharing enough information for the vendors to make an adequate decision is also critical.

Part VI
The Part of Tens

In this part . . .

I give you some handy reference material to help you keep on top of the latest developments in the RFID world. This part gives you information about the best Web sites, the most reputable equipment vendors, and the basics of the various RFID tag protocols.

The coolest chapter in this part is Chapter 20: I've gathered input from various experts throughout the RFID world and presented it in a short and sweet format that you can read through before you do your RFID setup and keep handy while you mount antennas and pull cable.

Chapter 18

Ten (Or So) Equipment Vendors

*W*hen you're looking for a partner or trying to get started with a technology, the toughest part is often either getting hold of the gear, particularly tags, or finding someone who is reputable. In this chapter, I show you some of the more reputable equipment companies and offer a bit of street knowledge about each one.

If you work with a vendor and don't get good service from the start, go to someone else as soon as possible. The companies best able to keep their customers happy are the ones with staying power. When you order hardware, tags, or services, demand a written service level agreement (SLA) with penalties for noncompliance. Although Chapter 17 focuses on outsourcing, many of the guidelines for finding a good outsourcing partner apply to RFID equipment vendors, too.

Alien Technology

Alien Technology Corporation, 18220 Butterfield Blvd., Morgan Hill, CA 95037; www.alientechnology.com; phone 408-782-3900

Alien is one of the leading companies in the passive RFID space. With significant venture capital funding, it has done an excellent job of marketing itself and achieving early market share. Alien hopes to be the premier provider of low-cost UHF (ultrahigh frequency) tags by using a manufacturing process called *Fluidic Self Assembly* to keep tag cost as low as possible. Alien was the original standard for Generation 1.0 EPC Class 1 tags. Its readers provide a wide range of options, from industrial readers for the supply chain to compact reader engines suited for hand-helds and printers. Many printer and hand-held manufacturers use Alien's reader board in their products.

Word on the street: Alien has the cheapest tags and the fastest order time on readers; however, its customer service and support model needs an overhaul. One issue is that Alien requires new customers to attend Alien academy — at a cost of $5,000 including a development kit — before they can get customer support.

ACCU-SORT

> ACCU-SORT Systems, Inc., 2800 Crystal Dr., Hatfield, PA 19440; www.accusort.com; phone 215-723-0981

For more than 30 years, ACCU-SORT has specialized in automatic identification technology, including barcode scanners vision systems. Recently, it has moved into RFID technologies.

The people at ACCU-SORT are print-and-apply specialists creating devices to print machine-readable information (like bar codes) on a label and applying it at high speeds. Their goal is to help clients streamline operations by managing materials and understanding data. They do not make RFID readers or tags; rather, they take other companies' hardware and incorporate it into their industrial equipment. The ACCU-SORT's FAST Tag RFID system incorporates barcode scanning, RFID labeling, RFID tag reading/writing, controls, and data management, including communications with the client's warehouse management system (WMS) or enterprise resource planning (ERP) system.

Word on the street: The company is building credible expertise around RFID and dedicating specific resources. They have a mobile solution that is perfect for most tag-and-ship types of solutions that take place in the warehouse. This mobile solution also includes control software, eliminating the need for a dedicated middleware piece. However, its control system, FAST Tag, is proprietary and works optimally as part of the total ACCU-SORT solution. If you're already an ACCU-SORT client, this isn't an issue. If you aren't, it appears you need to buy the whole system to get the maximum benefit.

Applied Wireless Identifications (AWID)

> Applied Wireless Identifications Group, Inc., 382 Route 59, Section 292, Monsey, NY 10952; www.awid.com; phone 845-369-8800

Applied Wireless Identifications, or AWID, has been in the wireless identification business since 1997. The initial focus was proximity cards for security systems and access control. In the early 2000s, Jeff Jacobsen, the former CEO of Alien Technologies, was hired as its new president. Since then, AWID moved to create an agile reader specifically for Wal-Mart suppliers. In early 2003, it released a self-contained very simple reader/antenna combination at a low price point. It has added power over Ethernet (PoE) (see Chapter 9 for details) and brought the lowest price on the market to the reader antenna combination.

Word on the street: AWID started off with a technology lead when Wal-Mart first released its mandate; however, it spent much of 2003–2004 toying with venture capitalists and lost its technology dominance and market position. It has a reader/antenna combination now at a street price of around $500, and can add additional antennas to make it even more attractive. AWID's funding indecision, however, has put it a step behind the other key providers. They are now trying to get back on their track of innovation.

FOX IV Technologies

FOX IV Technologies, Inc., 6011 Enterprise Dr., Export, PA 98203; www.foxiv.com; phone 724-387-3500

FOX IV has focused its print-and-apply systems on tough environments over the last 20 years. As a niche player, it has put smart engineers and designers to work solving challenging problems. To FOX IV, RFID is just another challenging problem. It was the first print-and-apply company to pass EPCglobal's certification for Class 1 and Class 0 tags in August of 2004.

Word on the street: FOX IV Technologies has proven itself to be a leader in tough situations and has taken some interesting approaches to the print-and-apply business. It tends to be more expensive than other print-and-apply vendors, however.

Impinj

Impinj, Inc., 501 N. 34th St., Seattle, WA 98103; www.impinj.com; phone 206-517-5300

Impinj is a highly innovative, venture-backed company that began life as a chip maker with innovative design. The folks at Impinj had a solution looking for a problem — RFID was made for them. They are providing integrated circuits (ICs) or chips for tag readers and also producing their own tags. Their competitive differentiation is the ability to power up a chip and backscatter a signal using lower power than traditional RFID tags. This makes the tags easier to read and helps create better-designed RFID networks because less reader power means fewer opportunities for RF noise. In addition, the Impinj engineers are designing readers for the generation 2.0 EPC protocol, which also keeps to their standard of innovation.

Word on the street: Impinj has a very talented and focused team, with strong partnerships among antenna and label manufacturers. Although their innovations have increased tag performance, the question is whether their prices will be competitive and whether this small company can keep up with production volume. The craftsmanship and price tag of a Ferrari will never be like an Oldsmobile, after all.

Intermec Technologies

Intermec Technologies Corporation, 6001 36th Ave. West, Everett, WA 98203; www.intermec.com; phone 425-348-2600

Intermec has been active in automatic identification technology (AIT) for decades. It has a significant embedded client base that uses its barcode technology. It also has a proprietary RFID system and is moving toward compliance with EPC and ISO (International Organization for Standardization) open standards, but has been slow to do so. Intermec claims existing deployments of RFID with manufacturing, food processing, security, and logistics in the United States; retail and logistics in Europe; biohazard waste management in Africa; and transportation systems in Japan. Intermec's UHF readers and tag systems are certified for operation in the United States at 915 MHz and 2.45 GHz, in Europe at 869 MHz (meeting narrowband requirements), and in Japan at 2.45 GHz.

Word on the street: When IBM decided to get out of RFID several years ago, Intermec purchased IBM's RFID intellectual property. Consequently, Intermec owns patents on everything from tag design to testing equipment. The company sat on the sidelines until the market matured before beginning to flex those patents. Intermec royalty demands were also partially responsible for slowing the process of an EPC protocol standard design. Thankfully they have not claimed royalty requirement for Gen 2.0 tags, but are indicating that they would require a royalty on Gen 2.0 reader design. Intermec's focus on previously developed intellectual property has left it behind the curve in terms of performance and capabilities in today's competitive marketplace, but their ownership of intellectual property has given them a strong lever to acquire or license other vendors' technologies. They should emerge a formidable provider in early 2006.

MARKEM

MARKEM Corporation, 150 Congress St., Keene, NH 03431;
www.markem.com; phone 603-352-1130

MARKEM, a marking technology company, has been in business for nearly
100 years. In late 2003, its scientists began experimenting with adapting their
existing print-and-apply solutions to RFID. They have developed several innovative solutions centered around print-and-apply at line production speeds.

Word on the street: MARKEM has allocated some of its top technical staff to
the problem of applying and encoding RFID tags at line speed. Although the
volume and scale of their production is unknown, the early indications of
performance are quite strong.

Symbol Technologies, Inc. (Formerly Matrics)

Symbol Technologies, One Symbol Plaza, Holtsville, NY 11742-1300;
www.symbol.com; phone 1-866-416-8545

Before its acquisition by Symbol Technologies in mid-2004, Matrics began as
an aggressive, venture-backed company with technology borne out of scientists from the National Security Administration (NSA). It designs and manu-
factures EPC-compliant RFID readers and tags, focused on a proprietary Class
0 protocol. It has patented a unique approach to assembling tags (known as
PICA) that will increase tag output and reduce overall cost, including using a
very fast-drying, inexpensive adhesive.

Word on the street: The Matrics/Symbol agile reader is clearly one of the
best readers out there. However, in 2004, some orders took up to eight weeks
to process because of issues with ramping up the contract manufacturing
house. Matrics/Symbol's AR-400 was the reader that Wal-Mart used in the first
phase of its RFID deployment in 2004. Matrics/Symbol tags, although only
Class 0, are well-designed, but large (giving better performance but less useful
placement options) and relatively more expensive than competitors' tags. All
in all, these tags are a great choice for closed-loop systems, which explains
their great success with airport luggage systems. With Symbol purchasing
Matrics, look for the company to provide integrated bar code and RFID systems and be quickly compliant with Gen 2.0 solutions.

ODIN technologies

ODIN technologies; 12120 Sunset Hills Rd., Suite 410; Reston, VA 20190; www.odintechnologies.com; phone 866-652-3052

ODIN technologies is the leading provider of specialized software and services for the RFID infrastructure. Starting with expertise from MIT's Auto-ID center and growing over the years with experienced physicists and RF engineers, ODIN has developed novel software to make the testing, deployment, and management of RFID networks highly successful. ODIN technologies was the first company to release software based on sound scientific methodologies for testing tag selection and placement. It has also created a system to design full RFID networks called FAS-D, the fully automated system design.

Because ODIN has had a background in the physics of RFID for many years, it's uniquely qualified to understand how to manage, monitor and keep RFID networks healthy. ODIN has the first monitoring and management system in the RFID industry which actually looks at tag read performance over time, using artificial intelligence, machine learning, and fuzzy logic in a patent-pending solution designed for keeping RFID networks up and working properly 24/7 in even the most remote locations.

Word on the street: ODIN technologies has worked with more of the Wal-Mart, Target, and Department of Defense clients than any other firm in the market-place on infrastructure testing, deployment, and monitoring. ODIN has a solid reputation for deploying 100-percent accurate solutions and is often the expertise behind many of the large consulting firms' efforts (who subcontract their work to ODIN). Its software products to test tags and monitor systems are the only ones on the market taking into account the complex physics of RFID, and are based on years of experience in the RFID space. ODIN also makes inexpensive and well-engineered RFID racks.

OMRON electronics

OMRON Regional Management Center; One East Commerce Drive, Schaumburg, IL 60173; www.omron.com; phone 847-884-0322

OMRON USA is a several hundred million dollar subsidiary of Japanese manufacturing automation giant OMRON — a $3 billion company. The company is well known for its manufacturing automation and control devices, ranging from optical label readers to high-speed fill-verification devices. OMRON has been creating RFID devices, mostly in high frequency (HF), for more than a decade and understands the physics of RF systems quite well.

OMRON is using its deep knowledge of industrial automation to develop cutting edge RFID solutions. The people in the U.S. division are spearheading the RFID efforts for the global organization. They have recently introduced UHF tags, readers, and antennas — all with proprietary technology and well researched design.

Word on the street: OMRON started its reader strategy by licensing the firmware and basic design from ThingMagic. Before production, it revamped the inner working of the reader and beefed up the components, creating an immediately optimized reader. OMRON's R&D division in Japan has also created very successful reader antennas and RFID tag designs and is looking to release its own reader design in 2005. The tag designs have a novel chip attachment process, using a crimping technology instead of adhesive, creating a much more durable tag. With support around the globe, a CEO who has taken personal interest in the RFID program, and great engineering with years of HF RFID experience, OMRON is a serious force to be reckoned with in the RFID marketplace.

SAMSys Technologies

SAMSys Technologies Inc., 2525 Meridian Parkway, Durham, NC 27713; www.samsys.com; phone 1-888-483-6646

A pioneer in multiprotocol, multifrequency readers, SAMSys has historically designed and deployed passive RFID readers (low frequency [LF], high frequency [HF], and ultrahigh frequency [UHF]) that support EPCglobal standards and other tag protocols and frequencies. Its innovative design flexibility has won significant contracts with both U.S. and European providers. SAMSys does not manufacture tags and has limited services, so all of its effort goes into making the best reader hardware possible.

Word on the street: For multiprotocol control and flexibility of data manipulation, the SAMSys reader is tough to beat. A savvy programmer can make the SAMSys reader work quicker and better than its competitors. However, the quality of the radio is slightly lower than some of the other readers, so if pure distance is an issue, SAMSys may lose out. It is also the most expensive reader on the market.

Texas Instruments (TI)

Texas Instruments Incorporated, 6550 Chase Oaks Blvd. Ms8470, Plano, TX 75023; www.ti.com/rfid; phone 1-888-937-6536

TI brands its RFID group, not surprisingly, TI-RFid (Texas Instruments Radio Frequency Identification Solutions). This group has significant experience in LF and HF and waited for the protocol dust to settle before venturing into UHF, but now it is going after the tag market with a vengeance.

Word on the street: TI has some very smart RFID folks working for it. Several of these folks have been involved with the various standards groups, which means they understand the direction the protocols are going and have a jump on designing the chips. TI will have some of the first Generation 2.0 tags available and has experience in high-capacity production to make sure they can meet demand. The only lingering question is whether TI will lose this fight to another competitor the way it lost the battle for the calculator to HP. With strong expertise in chips and RFID tags, TI is a good bet as an early winner. If TI focuses on more nimble innovation and cannibalizing its own products, it has a fighting chance to be a long-term winner.

ThingMagic

ThingMagic, 1 Broadway, Cambridge, MA 02142; www.thingmagic.com; phone 617-758-4136

ThingMagic is a small engineering start-up that was formed through early involvement with the Auto-ID Center at MIT. It was one of the first producers of an agile reader, and the first to use an Intel architecture coupled with a Linux core. Its latest reader, the Mercury IV, claims to be capable of reading all classes of tags and readers and has solid connectivity and very flexible controls and interfaces. Its top sales and marketing guy is Kevin Ashton, the former Procter & Gamble executive who was on loan to the Auto-ID Center as its Director and has been in the EPC space since it started.

Word on the street: ThingMagic has several very talented engineers and designers; however, its production capability and channel strategy is confusing to many end users. The good news for end users is that you can get the benefit of ThingMagic's technology but the reliable scalability and proven service of a large organization by going to companies to whom it has licensed technology. Tyco/ADT is one company with the basic ThingMagic design, or you can get an even better designed reader with some of the best antennas on the market and great global reach by going to OMRON, who also uses the basic ThingMagic technology foundation. Also noteworthy is the fact that a large independent reader manufacturer has licensed the technology as well.

Chapter 19

Ten Web Sites for Information on RFID

*N*avigating the choppy waters of the RFID Sea is bound to get you listing in your cubicle, so here are ten (or so) Web sites that will be your ballast. Keep in mind that this industry is changing quickly, new companies are being formed, older companies are being acquired, new products are being developed, protocols are evolving — well, you get the picture. The Internet can be as much a source of confusion as a source of information, but the following sites are bound to help keep you informed and up-to-date.

If you are considering using a consultant or outsourced firm, these sites are also a great way to keep them honest and assess their levels of knowledge. A little research will help you develop questions to ask a potential consultant and give you the ability to easily compare the answers with industry standards and expectations. If "trust but verify" worked during the Cold War, it most certainly will work for your RFID deployment.

RFID Journal Online

```
www.rfidjournal.com
```

If something that is only a few years old can be called the granddaddy of an industry, this is it. *RFID Journal Online* was started by Mark Roberti back in 2002, after a stint as managing editor of *Internet Week*. The site has all original content; it does not simply republish press releases or link to other sites. Some useful content about the RFID industry is always available. It is simple to navigate, has easy-to-understand headlines, and gives solid information. The only drawbacks are the amount of prime real estate given to ads and the logo that doesn't link back to the home page.

The RFID Journal Web site business model is a combination of successful advertising-driven revenue (you will notice a good number of ads) and subscription. Readers who choose not to subscribe can access the majority of the site and still glean a lot of useful industry information; subscribers get access to key feature articles and message boards. The journal also runs an annual conference called RFID Journal Live, and sponsors seminars and educational series. All in all, it's a great site for keeping up on the latest news in the industry. In fact, it's my home page at the office.

EPCglobal

```
www.epcglobalinc.org
```

The EPCglobal Web site is the "father" site to the regional EPC (electronic product code) organizations, like EPC US. You can get all the information about standards and technology at the site. However, much of the content is available for members only. This is particularly true of emerging standards and participation in workgroup-developed information. In order to browse the members-only sections, you need to subscribe to EPCglobal, which ranges from a few thousand dollars all the way up to $200,000, depending on the size of your organization.

The site is simple and easy to use, and the section that features position papers is particularly informative. The only ads you have to suffer through are about EPCglobal-specific events or programs.

For an introduction to EPC and EPCglobal's role in standards, see Chapter 2.

IDTechEx

`www.idtechex.com`

The IDTechEx is a United Kingdom site set up and run by a group of RF professionals who also do consulting and trade shows/conferences. There is a heavy push to buy their products and services, particularly the research and white papers. The site does a nice job, however, of giving a global picture to the RFID world, and addresses issues that are often overlooked in the United States.

The company has an extensive database of RFID case studies that is accessible only to subscribers. Subscription is several thousand dollars, however. The revenue model is subscription and sales of Web journals and studies. This makes for easy navigation because there are no distracting advertisements.

RFID Solutions Online

`www.rfidsolutionsonline.com`

RFID Solutions Online is part of the VertMarkets group of sites. The Vert people have spent the past five or six years refining how they offer sites that help businesses connect with solutions providers. The site is divided into a searchable vendor section and a general RFID information section. The vendor section allows you to find specific companies to meet your needs and to request a quote directly. The home page publishes vendor-developed Web sites that you can download for free — in exchange for your contact information. The site's editors are selective on the content they publish, so the information on the site is usually valuable, although sometimes it is a little on the salesy side.

RFID Exchange

`www.rfidexchange.com`

The RFID exchange is a newer Web site that is taking a comprehensive approach to data gathering. Like the IDTechEx site, it's a tool used to drive consulting business, but it doesn't sell research or white papers; rather, the site aggregates white papers and presentations from other sources. It has many white papers on the basics of RFID.

The site could benefit from a more professional design and usability makeover. For instance, when looking for a vendor, you have to browse through many Web pages, each of which contains only a handful of vendors listed in alphabetical order.

RFID Update

www.rfidupdate.com

RFID Update is a very user-friendly aggregation of weekly RFID stories. It also offers a subscription-based e-mail service. It has an active bulletin board and marketplace called RFID Talk that has real-world user issues and information.

The site makes its revenue from advertisers and has a large banner ad at the top of each page.

Auto-ID Labs

www.autoidlabs.org

The Auto-ID Labs' archive section is the best source for early research into the whole RFID/EPC movement. Here you can find great research papers and primers on the technology. A lot of the stuff is outdated and not so useful because the technology has changed so quickly, but some of the info is very helpful. This is a dot-org site and has no sponsorship or distracting ads to deal with.

Tom Scharfeld is currently part of the research team at Auto-ID Labs. He wrote his master's thesis at MIT on the physics principles of RFID specific to passive tags. This thesis is a great technical document for those wishing to understand more about the science of RFID. You can find it by searching for "Scharfeld" in the archives.

Auto-ID Lab @ Adelaide

http://autoidlab.eleceng.adelaide.edu.au/

The eccentric Peter Cole, probably the world's foremost expert on radio frequency identification, leads the University of Adelaide lab and he serves as the research Director for the Auto-ID Lab in Australia. The scientists at Auto-ID

Lab @ Adelaide have vast experience in the physics of RFID tags and readers and the development of RFID protocols. Peter is largely responsible for the EPC protocol and for helping the MIT team develop its technology.

The Resource section, where all sorts of good technical information is hidden, is the most useful section on the site.

The RFID Gazette

```
www.rfidgazette.org
```

Hear ye, hear ye; *The RFID Gazette* is printing all the news that's fit to be printed. This well organized aggregation site is designed around specific industry verticals — like pharmaceutical or automotive. Using more than moveable type, this publication does a nice job of providing a simple, easy-to-use interface for finding news articles specific to an area that may interest you. The categories, which you find on the navigation bar on the left, range from Privacy to Conferences. Another handy feature is the organization of archive articles by month, which makes so it easy to go back and find things you found interesting or worthwhile when you first read them. Although the lack of original content may be a drawback, the organization of the information is a plus.

UCLA's RFID@WINMEC site

```
www.wireless.ucla.edu/rfid/research
```

The University of California at Los Angeles (UCLA) has spawned an academic movement to challenge the electronic product code (EPC) created at MIT and to create a new middleware component. The site is the brainchild of UCLA professor Rajit Gadh. The content is minimal at this time, but it does hold the promise of an interesting alternative view of the RFID world. And there are no ads!

Slashdot

```
www.slashdot.org
```

And last but not least, this chapter wouldn't be worth its salt if it didn't pay homage to one of the greatest Web sites in the technology world: Slashdot.

Self-proclaimed "news for nerds," this site hones the sharp edge of the true geek in you and helps you keep up-to-date with all that is techie — from games to gamma rays. The site has great usability refined from years of user feedback and quick summaries of the major headlines.

The only drawbacks to the site are that the ads take up too much prime real estate on the right side of the screen, and there is no easy search location for finding out the Slashdot opinion on things.

Chapter 20

Ten Tips from the Experts

You know the old rule: Never buy a car from the first model year it was built. The idea is that a new car, from any company, is going to have a number of kinks to work out, and the only way to identify those issues is real-world usage. Prototyping and lab testing of a car will never tell you what a two-year-old child can do to the control knobs in the back seat, or how a bored English Setter can find entertainment in the cargo area, or that after ten thousand miles, that cheaper seal for the brake lights wasn't such a good idea. I could go on forever about gripes I've had with first-run cars, but you get the point. The same thing is true in an RFID system and network deployment. The first-run systems always have room for improvement.

In this chapter, I've collected what I think are the best tips out there from people who know. The handful of folks in this chapter have been doing RFID deployments for years and have made many mistakes and created some interesting shortcuts. I always say it's okay to learn from mistakes, as long as they are someone else's. Here's your chance to learn from other people's mistakes to make your RFID deployment easier, faster, and more effective.

Chris Fennig, ODIN technologies

Buy a spectrum analyzer *first*, before you do anything else. You can buy a good one dedicated for the frequency spectrum you're deploying (such as 902–928 MHz) for around the price of two readers. It's a great investment to keep you from guessing what's happening.

Get the right tag for your product. There are so many different types of passive RFID tags available now, with such a wide variance of price range, that it is critical to make sure that you pick the right tag for the job. The right tag is seldom the most expensive one, and you can save a significant amount of money by taking the up-front effort to do proper product and tag testing. Even if you farm out testing to an outside lab, the few thousand dollars to test your products could save hundreds of thousands in tag cost in the long run.

Antennas and RF energy are cheap; don't be stingy with them. The more diverse your antennas are, and the closer you can get them to the near field (that is, 33 centimeters away for UHF — see Chapter 4 for more details), the better your read success will be. Items that are RF unfriendly, such as cases of metal objects or liquids, also require more antennas than you might use. There's nothing wrong with putting four antennas on a conveyor instead of two to make sure you get more accuracy.

Over-engineer your read zone. If you need to read halfway across a dock door and it's five feet, set up the reader to read six or seven feet, as long as you aren't broadcasting RF energy to read past your desired boundary and cause potential interference with an adjacent zone.

Joe White, Symbol Technologies (Formerly Matrics, Inc.)

Choose the correct RFID technology that best suits your application because many options are available. Options include frequency selection — 125 KHz, 13.56 MHz (also known as HF), 860–960 MHz (also known as UHF), 2.45 MHz, and others, — regulatory environment, the type of product being tagged, the read range requirements, and RF environment.

Try to make sure you turn the transmit power of the radio only to a level that meets the application requirements and no more. It is tempting to always turn the power as high as it will go, but doing so may actually reduce performance due to RF noise interference or driving the radio into compression. (*Compression* is a state in which the RF noise floor exceeds what the radio receiver is designed to handle, thus rendering the radio unable to detect tag responses.)

Remember that UHF RFID has troubles with two types of products: liquids and metal. Liquids provide too much RF attenuation and prevent the energy from powering the tag, while metals ground the RF energy and prevent powering the tag.

Metal is not a good substance for passive UHF tags because it grounds the energy needed to power the tag. If you wish to tag metal items, try placing ¼ inch of cardboard or Styrofoam between the metal object and the tag.

To ensure best tag performance when tagging cases of liquid items, try placing the tag on the case where there are air gaps either between items or toward the top of the case. For example, when tagging a case of laundry detergent, the tag performs best on the case, near the top, by the handle of the detergent. This is where there's the least amount of liquid to interfere with the RF performance.

When tagging cases of products that prevent RF signal propagation, try placing the tag on the outside corner of the cases. In some instances, using two tags on the case at opposite corners will work best.

Duncan McCollum, Computer Sciences Corporation (CSC)

I think that the best way to ensure RFID project success is to focus on one problem and solve that well. After you have that first project success under your belt — no matter how small — you can easily transfer what you learned to the next project.

Too often, RFID projects fail because we try to solve all our problems with RFID. Find a problem in your organization that is well defined, within your control, and measurable. If RFID is the best solution for that problem (that's another way to fail, RFID is not the solution to every problem!), start small, master the fundamentals of the technology, get an early win, and showcase your solution to the rest of the organization.

A good example is tracking full pallets in and out of a distribution center. Often, because a pallet in and pallet out scenario does not go through a sort line where there are bar codes reading each case, pallets going through this cross-docking scenario are often not counted as accurately as broken-down pallets. This would be a great first case for RFID: All full pallets going in and out without being broken down could go through an in-bound and out-bound RFID portal.

It's important to remember that we are dealing with a technology that is developing at a very fast rate; the prudent course is to stay just behind the edge of the adoption curve. That doesn't mean doing nothing; that means getting involved, testing, and understanding so when you've got the right solution you can move quickly — that's competitive advantage!

Dr. Daniel Engels, MIT Auto-ID Labs

People get into trouble when they set up two readers adjacent to each other (on dock doors for instance) and the transmit antennas from each reader are facing receiving antennas from the other reader. This creates something called compression. By *canting* the antennas (making sure you know which antennas are transmit and receive and positioning them with as much distance between them as possible), you avoid compression problems and increase read performance.

If you're having trouble reading tags, a good way to troubleshoot is to determine if it is a forward link issue (lack of power to the tag) or return link issue (the tag is responding but not being heard). By separating the transmit antennas from the receiving antennas, you can diagnose this problem. You need a reader that has a dedicated transmit (Tx) port and a dedicated receive (Rx) port.

- **Forward link issues** usually occur when there is a lack of power to the tag due to signal attenuation, insufficient RF power, or poor quality tags are being used. Try moving the transmit antenna close to the tag while leaving the receive antenna stationary until there is a successful read. Forward link problems are best solved by adjusting the power up, moving the antennas closer, trying a more sensitive RFID tag, or providing for more antenna diversity (antennas covering more of the area in the proposed RFID interrogation zone — by going from two to four antennas, for instance).

- **Return link limitations** usually occur when the tag is getting enough power (above –10 dBm) but the receiver is unable to hear the tag's backscatter (response). The following table lists the possible causes of return link limitations and their solutions.

Cause	What It Means	Possible Solution
Ambient electromagnetic noise (AEN)	The surrounding RF environment is drowning out the weak signal coming back to the reader.	Always make sure you do a Full Faraday Cycle Analysis to see what the RF environment looks like before deploying your system. Chapter 7 explains how you do this analysis.
Receiver compression	The noise of the transmit signal is overpowering the sensitivity of the radio receiver and basically muting the signal.	Try moving the offending transmitter away from the receiver.

Cause	What It Means	Possible Solution
Tag attenuation	The return signal of the tag is too low for the receiver to pick up the tag.	Try moving your receiver closer to the tag until it begins reading, and make sure that you are using the right tag and placement by going through proper tag testing. See Chapter 8 for all the skinny on tag testing.

Dr. Patrick King, Michelin Tire Corporation

Make sure you create read zones with known outer boundaries. This will help you avoid issues with phantom or ghost reads from other areas. The easiest way of designing an outer boundary edge without a spectrum analyzer and signal generator is to put the tag you are using on an empty cardboard box and place it just past where you want to be able to read items. Turn down the power on the reader until there are no more reads. Because of a phenomenon called *hysteris,* you should also start the reader at its lowest power and turn it up incrementally until you can read the tag. The two power levels should match. If they don't, go with the lower level.

In a dock door environment, try to angle the antennas slightly outward. Most antennas will be mounted just inside the dock door. Angling them to the outside will help avoid interference with adjacent doors.

To capture RFID tag reads in the order that they occur in a conveyor application, angle the antennas on a 30- to 45-degree angle in the flow of the conveyor. The middleware application that you use should allow you to set parameters so that when a tag enters the field and is first read, it can be recorded in that order in your application.

Steve Kowalke, ACCU-SORT Systems

For your first mandated implementation, don't go overboard — unless you have a great understanding of the technology. There's no need to do a full implementation, spend several hundred thousand dollars on middleware and readers, and then find out after you are generating data that you've taken the wrong approach.

The best first step into the RFID foray is to do a simple solution. An easy way that we at ACCU-SORT have solved the problem is to provide a mobile solution in the distribution center. This solution works when the pallet is going out to Wal-Mart or Target. A dock worker can scan the bar code on each case, and the mobile RFID station writes and records the tag and the EPC number. A solution like this, and a verification portal, can be installed in a matter of a month after completing a Full Faraday Cycle Analysis and proper SKU testing. All in all, a compliance solution like this can cost under $75,000 and give you a good first taste of what kind and how much data your system will generate.

Team Tag-IT, Texas Instruments

Be mindful of FCC requirements, particularly if you are an early adopter of the technology. Ask your vendors about FCC certification early in the process — what type of certification do you need (such as a site license or experimental license), if any, and how will you acquire it.

Keep FCC compliance in mind, too. We were tasked with installing a one-off antenna system. There exists an off-the-shelf reader/antenna combination on the market that has FCC approval and would probably work well for our niche install. However, the price was a few hundred dollars more than we budgeted to spend, so we decided to manufacture the antenna inhouse. It worked exactly like the competitive model. But, after install, we received a quote on FCC approval testing for just under $10,000. So much for that ROI.

Liberally resource your internal manpower early in the program. Employees involved from the start often champion the program within the organization.

Have a change management procedure that communicates to your team changes in the schedule, technology, scope, personnel, and specifications.

Know what the long lead items (equipment or label procurement, site preparation, and so on) are as early as possible in the deployment. This minimizes impact to your project plan.

As they say in the Boy Scouts: Be Prepared.

Kevin MacDonald, Lead RFID Architect, Sun Microsystems

Make sure there is a dedicated project manager from both sides. Your internal team should have a single point of contact who knows your legacy architecture and has a good grasp of your technology road map, and your RFID vendor should have one person who can coordinate the physics, the networking, the middleware design, and application integration.

Set up an RFID dashboard to avoid data overload. This is something as simple as an HTML or PHP user interface that gives information back on your RFID network. The dashboard could be set up to aggregate only pertinent information from each RFID interrogation zone, each distribution center, or some other collection of the RFID network. RFID can create a lot of erroneous data if you start with bad data collection. Some manufacturers have stated that one SKU tracked through seven locations can generate 4 tetrabytes of data per year. The reality is, in the end, that the useful, relevant data is a very small amount. For example, if you're tracking the movement of a widget through the supply chain, you may only want to know what average times are in a particular location and what cases took extremely long times. That will allow you to drill down into the process and figure out exactly what was going wrong and, more importantly, what can be fixed.

The way to set up the dashboard is to take a dual approach of modeling the data and creating sample reports. To get a model to work from, you need to gather data coming from a sample slice of your operation for a period of time that is long enough for problems to arise. If it's a very high-volume operation, this may be a week or two; if it's lower volume, it may be only a couple of days. From there, you can run sample reports to determine where useful data may reside and how you can adjust your filtering so that the bandwidth and edge data requirements don't become overwhelming.

Mark Nelson, Savi Technology

There are five critical players in bringing an RFID deployment from a pilot to a full-blown commercial run. Those five players are

- ✔ **Analysis Coordinator:** One person to work on a part-time basis with the Pre-Run Analysis Team for requirements gathering

- ✔ **Project Coordinator:** One person to work as an operational contact during the commercial run as an internal project coordinator

- ✔ **Participant Solicitation:** Encourages key supply chain partners such as shipping lines, LSP, and factories and warehouses to participate in production deployment

- ✔ **User Feedback Committee:** Provides on-going feedback from user group on an ad-hoc basis

- ✔ **Executive Review:** Executive team members to set a review meeting at conclusion for results presentation

In Chapter 12, you can find more details about working through a pilot deployment to a production deployment. Chapter 16 offers more about putting together an RFID team and making the business case to executives.

Chapter 21

Ten (Or So) RFID Standards and Protocols

* *

* *

*M*any people are touting RFID as a replacement for the bar code, which is universally accepted and understood to have a standard. However, this isn't quite the truth: There are more than 200 different bar code standards. Similarly, although movement toward universally accepted standards is helping RFID gain ground on the bar code's turf, you'll find a variety of RFID standards for specific applications, such as animal tracking or access control cards. These industry-specific applications lead vendors to develop RFID as the market demands them, without any concern for interoperability with other niche applications.

The emerging global supply chain and collaboration among international partners has led to attempts to create one global standard within an RFID network. Many issues remain, but the evolution is progressing much quicker than the bar code industry did.

Although the bar code is still the most prevalent Auto-ID technology, the evolution to an advanced replacement — RFID — has put the focus on standards and protocols. In Chapter 2, I explain why RFID is the new King of the Hill in Auto-ID technologies. In this chapter, I show you some of the standards for Auto-ID in order to help you keep tabs on your business needs or standards you may want to become involved in and help shape.

EAN.UCC

EAN International and the Uniform Code Council (UCC), which is a member organization of the EAN International, are responsible for the co-management of the EAN.UCC System and the Global Standard Management Process (GSMP). The EAN.UCC System maintains and sets standards for bar codes, electronic data interchange (EDI) transactions sets, XML schemas, and other supply-chain solutions. By administering the assignment of company prefixes and coordinating the accompanying standards, EAN International and the UCC claims to "maintain the most robust item identification system in the world."

Although the main role of EAN.UCC is the assignment of bar codes and unique numbers for enterprise products and locations, the EAN.UCC also has two initiatives for "tracking and trace" and "data synchronization." The GTAG (Global Tag) initiative was launched jointly by EAN International and the UCC in 2000 for tracking and tracing items. The GTAG initiative was designed to allow for both short-term solutions and a marrying of business processes with standards development. These standards cover both UHF RFID technology (including air interface) and data structure. The air interface efforts of GTAG have become part of another standard — ISO 18000-6.

Go to www.ean-ucc.org to get the lowdown on this group.

EPCglobal

EPCglobal is leading the development of industry-driven standards for the electronic product code (EPC) to support the use of radio frequency identification. EPCglobal is responsible for assigning blocks of numbers to its members, and also has defined the EPC tag protocol, which covers data structure written to the tag and the air-interface protocol. They have also worked to develop a data exchange and dubbed it the EPC network. In addition to providing unique numbers for product definition, this organization has also defined a framework for implementing an RFID solution. This framework defines everything from the specifications of the RFID tag to the interfaces that relate a unique number to an enterprise. This requires the definition of repository functions and interfaces to devices and back-end systems. The purpose of the EPCglobal framework is to provide the ability to "track and trace" products through the supply chain. Find more information about EPCglobal at www.epcglobal.com.

UCCnet

A subsidiary of the Uniform Code Council (UCC, a global standards organization), UCCnet is a nonprofit organization providing an electronic foundation for implementing industry-developed standards. UCCnet stores a registry, or catalog, of item-level data and provides an engine that synchronizes data between businesses using the UCCnet platform.

Companies called *data pools* provide connection and access to the GLOBAL registry. For a company to be its own data pool, it must be certified by the UCC. If your firm is like most companies, you'll send data to a data pool partner, which then connects to the UCCnet network. This is very similar to Value Added Networks (VANs) and their relationships in EDI.

UCCnet's purpose is to synchronize data between enterprises. Whereas the EPCglobal organization maintains a registry of product numbers, UCCnet maintains the details associated with those numbers. As I mention in Chapter 2, you can't tell much about a product by simply knowing its EPC number. Most retailers will eventually require suppliers to provide this detailed data so that retailers can validate goods received, automatically update inventories, and reduce errors in invoices and payments.

The bottom line to suppliers is that if you are working with one of the giant retailers, such as Wal-Mart, you need to adopt UCCnet to collaborate within the supply chain. RFID data won't likely be sent on its own dedicated network; rather, RFID systems will leverage existing technology already widely used by trading partners. Visit UCCnet at www.uccnet.org.

ISO/IEC JT1/SC17

The International Organization for Standardization (ISO) has partnered with the International Electrotechnical Commission (IEC) to address standards for identification cards and related devices. Joint Technical Committee 1, Subcommittee 17 is working on this. SC17 includes the standards 10536, 15693, and 14443. Table 21-1 outlines what these standards address.

Table 21-1	ISO/IEC Identification Card Standards	
Standard	*Type of ID Card*	*What Does Each Part Cover?*
ISO/IEC 10536 Identification cards — contactless integrated circuit(s) cards	Smart identification cards, using RFID at 13.56 MHz	**Part 1:** Physical characteristics **Part 2:** Dimensions and location of coupling areas **Part 3:** Electronic signals and reset procedures **Part 4:** Answer to reset and transmission protocols
ISO/IEC 14443 Identification cards — Proximity integrated circuit(s) cards	Smart identification cards with a longer range (up to 1 meter), using RFID at 13.56 MHz	**Part 1:** Physical characteristics **Part 2:** Air interface **Part 3:** Initialization and anticollision **Part 4:** Transmission protocol

(continued)

Table 21-1 *(continued)*

Standard	Type of ID Card	What Does Each Part Cover?
ISO/IEC 15693 Contactless integrated circuit(s) cards — Vicinity cards		**Part 1:** Physical characteristics **Part 2:** Air interface and initialization **Part 3:** Anticollision and transmission protocol

ISO/IEC JTC1/SC31/WG4

ISO is also working with the IEC to focus on automatic identification and data-capture techniques. JTC1/SC31 (Joint Technical Committee 1, Subcommittee 31) is focusing on data capture. The big enchilada of this committee is the use of RFID for item management, and the committee is focusing on the standards 15961, 15962, 15963, and 18000:

- ✔ **ISO/IEC 15961 RFID for Item Management — Data Protocol: Application interface:** This standard addresses the common functional commands and syntax features (for example, RFID tag-types, data-storage formats, and compression technology). The standards for the air interface protocols don't affect standard 15961. Rather, 15961 goes hand-in-hand with standard 15962, which addresses the overall data-handling protocol.

- ✔ **ISO/IEC 15962 RFID for Item Management — Protocol: Data encoding rules and logical memory functions:** This standard addresses the interface procedure that an RFID system uses to exchange information for item management. It creates uniform and correct formatting of data, the structure of commands, and the processing of errors.

- ✔ **ISO/IEC 15963 RFID for Item Management — Unique Identification of RF Tag:** This standard addresses the numbering system, the registration procedure, and the use of RFID tags. The standard is designed to address quality control during the manufacturing process. It also addresses the traceability of the RF tags during their manufacture and usable lifetimes and the anticollision issues of having many tags in an interrogation zone.

- ✔ **ISO/IEC 18000 RFID Air Interface Standards:** ISO standard 18000 is designed to create true global interoperability in how tags and readers communicate, even at different frequencies. The goal of this standard is to ensure that the air interface protocol is universal.

18000 has seven different parts. Part 1 is the system architecture for RFID for item management, including integration with legacy systems and interoperability. Parts 3 and 6 are the two critical sections of 18000. Part 3 enumerates two different modes of operation, which are not interoperable although they are designed not to contend/interfere with each other. Mode 1 is based on ISO 15693 with improvements, and Mode 2 lays out a new high-speed communication option. Part 6 also defines two kinds of operation, known as types A and B.

For more details on these standards and to stay up with the latest, you can check www.iso.org for more information.

The ISO/IEC 18000 series of standards differ from the EPC standard and actually coexists with it because the ISO standards deal with only the air interface protocol and not the numbering structure or the physical implementation of the tags and readers.

AIAG

The Automotive Industry Action Group (AIAG) is a professional organization working on supply-chain issues in the automotive industry. Its mission is to propose industrial standards and provide guidelines to the automotive industry. The RFID data standard that the AIAG defined for tires is called the B-11 standard. It provides information about the manufacturer, tire size, type (including Department of Transportation data), and additional optional information. Because it has more information than just an identification number, the tag must be able to carry more data that is currently defined by EPCglobal. Check out www.aiag.org for more information.

Container Shipments

Since the terrorist attacks of September 11, 2001, the U.S. government has been concerned about the vulnerability of American ports. As a result, two government initiatives ensure the safety of shipments arriving in U.S. ports: the Container Security Initiative (CSI) and Smart and Secure Tradelanes (SST).

Container Security Initiative (CSI)

CSI was announced in January 2002, and so far has been implemented at the top 20 foreign ports whose exports collectively represent approximately two-thirds of the volume of containers imported into the United States. The governments of these ports have already agreed to implement CSI.

CSI equips ports with four core capabilities that improve security:

- Establish security criteria for identifying high-risk containers based on advance information.
- Prescreen containers at the earliest possible point.

> ✔ **Use technology to quickly prescreen high-risk containers.**
>
> ✔ **Develop secure and "smart" containers.**
>
> This fourth point is where RFID comes into play and where the CSI intersects another government initiative: Smart and Secure Tradelanes (SST).

You can find more about CSI at `www.customs.gov/xp/cgov/enforcement/international_activities/csi/`.

Smart and Secure Tradelanes

Industry is driving SST, which uses an open technology platform. These industries are working with U.S. Customs, the Transportation Security Administration, Operation Safe Commerce, C-TPAT (Customs-Trade Partnership Against Terrorism), and the Container Security Administration in order to improve the security and productivity of cargo shipments.

The RFID-enabled seals that SST employs let you know *automatically* when a cargo shipment has been tampered with by going "silent." For this system to work, all container seals in a given area are constantly logged and monitored: If any seal suddenly stops responding, it can be flagged for action. Because counterfeiting these seals is impossible, no one can remove a tag and replace it with another.

Active RFID tag manufacturer Savi Technology (`www.savi.com`) developed the SST technology with a grant from the Department of Defense. Since receiving clearance to market the technology for civilian use, Savi has been actively building support for the product within the container transport community. For more information, see `www.savi.com/products/casestudies/wp.sst_initiative.pdf`.

Appendix

Glossary of Electrical, Magnetic, and Other Scientific Terms

• •

Absorption: A certain type of optical attenuation that converts energy waves (such as RF waves) into a different form of energy (such as heat) by interacting with the material properties of an object.

AEN: *See* Ambient electromagnetic noise.

ALOHA slot: This data transmission protocol was developed for one of the early local area networks. Hailing from the University of Hawaii, it describes a method of transmitting data that avoids data collision.

AM: *See* Amplitude modulation.

Ambient electromagnetic noise (AEN): Radio signals present in a specific location that need to be identified in order to determine the level of potential interference with an RFID system.

American Standards Code for Information Interchange (ASCII): A binary, 7-bit data encoding scheme comprising 128 codes used to represent all of the upper- and lowercase Latin letters, numbers, punctuation, and so on.

Amplifier: This electronic device increases a signal's transmitted power. You typically use it in-line between a reader and an antenna if the antenna needs a stronger signal.

Amplitude: The height of a wave.

Amplitude modulation (AM): A signal encoding technique that modifies the amplitude of a radio wave (as opposed to its frequency) depending on the behavior of the signal that it is transporting. By changing the amplitude of a wave, an RFID tag can transmit a binary signal to the reader. The method of changing the amplitude is known as amplitude shift keying, or ASK.

Analog signal: A signal that transmits continuous data, rather than discrete data (like a digital signal).

Antenna gain: The ratio, usually expressed in decibels, of the power needed for an antenna to produce the same field strength in a specific direction. The higher the gain, the more powerful the energy output. Antennas with higher gain will be able to read tags from farther away.

API: Application programming interface.

Application-specific integrated circuit (ASIC): A specially designed chip for a specific purpose — like RFID.

ASCII: *See* American Standards Code for Information Interchange.

ASIC: *See* Application-specific integrated circuit.

Asynchronous: Data that travels in a different time sequence (as in not back and forth at the same time) or directional properties and does not share a common clock source.

Attenuation: A reduction of the signal strength that is usually caused by some other factor. For instance, liquid can attenuate the signal of an RFID tag so that it can not send back a strong enough signal for the reader to read the tag.

Backscatter: The method of collecting inbound energy, changing the properties of that energy, and then reflecting it back to a receiving device (like an RFID reader). Most passive RFID tags use backscatter technology to communicate.

Bandwidth: The number of bits per second transmitted across a channel (for example, a T-1 transmits at 1.54 Mbps); or the range of frequencies a transmission system operates within (such as 902–928 MHz).

Baud: The signaling speed of a communication device. This is sometimes used synonymously with, and is usually equivalent to, bits per second.

Bend radius: The maximum amount a cable can be bent before serious signal attenuation occurs. This is usually a functional characteristic of cable and fibre. It's important to know the bend radius as you wire up RFID antennas.

Bidirectional: A system that is capable of transmitting simultaneously both forward and backward.

Bin: A value in binary notation.

Binary: A Base 2 counting scheme that has 0 and 1 as its primary digits. Also known as *digital*.

Bit rate: Bits per second.

Bluetooth: An open wireless standard operating at a transmission speed of 1 Mbps that is used for short-range transmissions.

Bus: The parallel device that interconnects the components of a computer.

Capacitance: The ability, measured in farads, to store an electrical charge in a circuit.

Capacitive coupling: The transfer of electromagnetic energy from one circuit to another through mutual capacitance. This is important in RFID because a tag, after it's given a charge from a reader, couples with that reader and stores that charge for a certain period.

Carrier frequency: The main frequency of a transmitter or RFID reader. This frequency is modulated in order to transmit information.

Carrier Sense, Multiple Access with Collision Detection (CSMA/CD): The access scheme used in Ethernet LANs, which formed the basis for many of the early slot protocols used in RFID. A device that is ready to transmit data first checks the channel for other data being transmitted. If the channel is clear, it transmits its data. If two devices transmit simultaneously, a collision occurs, and both devices abort the data transmission and then wait a random amount of time before transmitting again.

Checksum: A small bit of code added to a data block to verify its complete transmission. Often, an RFID microchip has a checksum that is checked before and after data is transmitted from the tag to the reader to determine whether the data has been corrupted or lost. One type of checksum is the *cyclic redundancy check* (CRC).

Chip: *See* Integrated circuit or chip.

Circularly polarized antenna: A UHF reader antenna that broadcasts RF waves in a circular pattern. Because of its design, it is less orientation-sensitive — you don't have to make sure the tag is in a perfect position in order to be read — but it has a shorter read range than a linearly polarized antenna, which *is* orientation-sensitive.

Close coupling smart card: A card that is similar to contact-based smart cards and is characterized by extremely short read ranges.

Complementary Metal Oxide Semiconductor (CMOS): A form of integrated circuit (IC) technology typically used in low-speed, low-power applications.

Compression: The process of eliminating redundant information from data in order to reduce the size of transmitted data without losing data integrity.

CRC: *See* Cyclic redundancy check.

CSMA/CD: *See* Carrier Sense, Multiple Access with Collision Detection.

Cyclic redundancy check (CRC): A type of checksum function using a mathematical technique for verifying the integrity of a data file to make sure that no data has been lost or corrupted. In the EPC data structure, a CRC is returned as the last block of data signifying a complete and successful read.

Decibel (dB): A logarithmic measure of the strength of a transmitted signal, most commonly associated with sound waves, but used to measure various types of waveform communication (like RF). Because it is logarithmic, a 20 dB loss indicates that the received signal is one one-hundredth its original strength.

Die: The silicon block onto which circuits are etched to create a chip.

Dielectric: A substance that is nonconductive, but that can sustain an electrical field.

Diode: A semiconductor device that allows current to flow only in a single direction.

DTD: Document Type Definition.

EBCDIC: *See* Extended Binary Coded Decimal Interchange Code.

Edge: The outer bound or periphery of any network. This is where functions such as aggregation, quality of service, and Internet protocol take place. The information and intelligence of a network most often is performed and stored in "edge devices."

EEPROM (Electrically Erasable Programmable Read-Only Memory): A type of data storage specifically for microchips in which bytes can be individually erased and reprogrammed. RFID chips are available with EEPROM and therefore can be programmed with specific numbers (the bytes changed) in the field.

Effective isotropic radiated power (EIRP): A measurement, usually expressed in watts, of the output of an RFID reader antenna. *See also* Effective radiated power (ERP).

Effective radiated power (ERP): A measurement, usually expressed in watts, of the output of RFID reader antennas used in Europe and elsewhere. *See also* Effective isotropic radiated power (EIRP).

Electromagnetic interference (EMI): Distortion caused when the radio waves from one device, such as a cell phone or overhead lighting, interfere with the radio waves from another device, such as an RFID tag reader.

Electronic article surveillance (EAS): Loss-prevention technology using passive RFID surveillance. This surveillance uses simple electronic tags that can be turned on or off. When an item is purchased at a store or checked out from a library, the tag is turned off during checkout. If a tag hasn't been turned off, an alarm sounds when the item is carried through a gate area. EAS tags have been widely used in the retail and pharmaceutical world for the past twenty years.

Electronic product code (EPC): Both a numbering scheme for storing data on an RFID tag (the EPC number) and a protocol for data communication and data storage (the EPC protocol). The EPC number is the equivalent to and eventual replacement for UPC barcodes for most general supply chain applications.

EMI: *See* Electromagnetic interference.

Enterprise resource planning (ERP): A technique for managing customer interactions through data mining, knowledge management, and customer relationships management (CRM).

EPC: *See* Electronic product code.

ERP: *See* Effective radiated power; Enterprise resource planning.

European Telecommunications Standards Institute (ETSI): The European Union body that recommends telecommunications standards for adoption by its member countries.

Evanescent wave: Light that travels down the inner layer of the cladding rather than down the fibre core of an optic cable.

Extended Binary Coded Decimal Interchange Code (EBCDIC): An 8-bit data-encoding scheme used to express alphanumeric characters.

Faraday Effect: Also known as the *magneto-optical effect,* it explains the extent to which materials can cause a linearly polarized signal to change when placed within a magnetic field that is parallel to the direction of the antenna.

Far-field communication: Communication between an RFID tag and reader that occurs beyond the distance of one full wavelength from the reader, typically with UHF and microwave systems. The far-field signal decays as the square of the distance from the antenna. *See also* Near-field communication.

FM: *See* Frequency modulation.

Forward Error Correction (FEC): An error-correction technique in which enough additional information is sent along with transmitted data that a receiver can both detect an error and actually fix it without requesting a resend.

Frequency-agile: The ability of a receiving or transmitting device to change its frequency in order to take advantage of alternate channels.

Frequency hopping: A transmitting device's continual shifting from frequency to frequency within a specific bandwidth range in order to comply with the requirements by certain standards bodies not to transmit on any one unlicensed frequency band for more than a certain length of time. In the United States, this is governed by the FCC Rules Part 15.

Frequency modulation (FM): A signal-encoding technique in which the frequency of an electromagnetic wave is modulated according to the behavior of the signal that it is transporting.

Full-duplex: Two-way simultaneous transmission.

Full Faraday Cycle Analysis: A way of gathering time-dependent spectrum analysis data across a specific band of operation to determine the effects of various ambient electronic noise on an operating environment deploying radio frequency.

Gigabit Ethernet: A version of Ethernet that operates at 1,000 Mbps.

GTIN: Global Trade Identification Number.

Hex: A value in hexadecimal notation.

High-frequency (HF): The frequency bandwidth from 3 MHz to 30 MHz. HF RFID tags typically operate at 13.56 MHz, can normally be read from less than 3 feet away, and transmit data faster than low-frequency tags, although they consume more power than low-frequency tags.

Impedance: Impedance is a measure of resistance to electrical current when a voltage is moved across it. This resistance is the combined effect of capacitance, inductance, and resistance on a signal, measured in ohms (Ω), and representing the ratio of voltage to the flow of current allowed. According to Ohm's law, voltage is the product of current and resistance at a given frequency.

Inductance: The property of an electric circuit by which a current flowing through it is changed, measured in henrys (H) and represented by the character *L*.

Inductive coupling: The shifting of electromagnetic energy from one circuit to another by virtue of the inductance between the two circuits. Inductive coupling may be deliberate and desired, as with an impedance matcher that matches the impedance of a transmitter to an antenna to guarantee maximum transfer of power, or it may be unplanned and unwanted.

Industrial, Scientific, and Medical (ISM) bands: The group of unlicensed, unregulated frequencies of the electromagnetic spectrum consisting of the 902–928 MHz range in the United States, or the 864–870 MHz range in Europe.

Inlay: An RFID microchip connected to an antenna and mounted on a substrate. Inlays are essentially unfinished RFID labels that are sold to label converters who turn them into smart labels.

Integrated circuit or chip (IC): A microelectronic semiconductor device made up of interconnected transistors and other electronic components. Most RFID tags use ICs.

Interrogation zone: The area where an RFID tag can be powered up and read, often between an array of antennas.

ISM: *See* Industrial, Scientific, and Medical bands.

Lambda (λ): The representation for a single wavelength in electromagnetic calculations.

Linear-polarized antenna: A UHF antenna that focuses the radio energy from the RFID reader in a narrow beam, increasing the feasible read distance and providing greater penetration through dense materials. Tags designed for use with a linear-polarized antenna must be aligned with the reader antenna in order to be read correctly.

Loss: The reduction in signal strength occurring over distance, expressed in decibels (dB).

Low-frequency (LF): The frequency bandwidth from 30 kHz to 300 kHz. Low-frequency RFID tags typical operate at 125 kHz or 134 kHz. Low-frequency RFID tags have to be read from within three feet, and their data transfer rate is slow, but they are less susceptible to interference than UHF tags.

LSbit: Least significant bit.

LSbyte: Least significant byte.

Manchester encoding: A data transmission code in which data and clock signals are combined to form a self-synchronizing data stream.

Material dispersion: A dispersion effect caused by the fact that different wavelength signals travel at different speeds through a medium.

MIB: Management Information Base.

Microwave tags: A term that is sometimes used to refer to RFID tags that operate at 5.8 GHz. These tags have a very high transfer rate and can be read from as far as 30 feet away, but they are more expensive and consume more power than other types of RFID tags. (Some people refer to any tag that operates above 415 MHz as a microwave tag.)

Modulation: The changing of a carrier wave to cause it to carry information. *See also* Amplitude modulation; Frequency modulation; Phase modulation.

MSbit: Most significant bit.

MSbyte: Most significant byte.

Mutual inductance: The tendency of a change in the current of one coil to affect the current and voltage of a second coil, resulting in the production of voltage. The voltage always opposes the change in the magnetic field produced by the coupled coil.

Near-field communication: Communication between an RFID tag and reader that occurs within one full wavelength of the reader, typically with low- and high-frequency systems. The near-field signal decays as the cube of the distance from the antenna. *See also* Far-field communication.

Noise: Unwanted ambient electrical signals that cause unpredictable impairment to RFID communications.

Overhead: The part of a transmission stream that a network uses to control and direct the data to its destination.

Packet: A data-carrying entity carried inside a frame or cell; a packet is also called a *datagram*.

Patch antenna: A small, square RFID reader antenna made from a solid piece of metal or foil.

Path loss: The degradation of an RFID signal based on the inverse square law for electromagnetic signals, caused by the concentration of the signal being spread out over a greater receiving area.

Penetration: The ability of a radio frequency to pass through nonmetallic materials. Low-frequency systems have better penetration than UHF systems.

Phantom read: The report of the presence of a tag that doesn't exist in its designated interrogation zone. Also called a *phantom transaction* or *false read*.

Phase modulation: The process of modulating the phase of an electromagnetic wave so that it can carry information.

Photon: A fundamental unit of light that has energy and momentum, but no mass or electrical charge. Sometimes referred to as a *quantum of electromagnetic energy*.

Polarization: The principle of changing the direction of the magnetic field, used to shape the propagation or interrogation zone.

Protocol: A set of rules that aids communications.

Proximity-coupling smart card: A card designed to be readable 10 to 15 centimeters from the reader.

Quiet zone: The blank areas on either side of the Universal Product Code (UPC).

Refraction: The shift in direction that occurs as a light wave passes from one medium to another.

Refractive index: A measure of the speed at which light travels through a medium.

Repeater (or Regenerator): A device that re-creates a signal and passes it on to another destination at a greater strength.

SAW: *See* Surface Acoustic Wave.

SGLN: Serialized Global Location Number.

Simplex: One-way transmission.

Slotted ALOHA: A variation on the ALOHA protocol in which stations transmit at predetermined times to ensure maximum throughput and minimal collisions. *See also* ALOHA slot.

SOF: Start of frame.

SSCC: Serial Shipping Container Code.

Standards: The published rules that govern an industry's activities.

Surface Acoustic Wave (SAW): A technology used for communication employing low-power microwave radio signals which are transformed into ultrasonic acoustic signals by a voltage-producing crystalline material in a transponder. Variations in the reflected signal can be used to provide a unique identity.

Synchronous: A term that means that both communicating devices derive their synchronization signal from the same source.

Tarred: What people down south get after reading long glossaries.

TBD: To be determined.

Transceiver: A device that incorporates both a transmitter and a receiver in the same housing.

Transponder: A device that comprises a transmitter, a receiver, and a multi-plexer in one; or a device that receives and transmits radio signals at a prede-termined frequency range. After receiving a signal, a transponder rebroadcasts it at a different frequency. An RFID tag is a classic example of a transponder.

UID: *See* Unique identifier.

Ultrahigh frequency (UHF): The frequency bandwidth from 300 MHz to 3 GHz. Normally, RFID tags that operate between 866 MHz and 960 MHz can send information faster and farther than high- and low-frequency tags. UHF signals can't pass through many items with a high water content, such as fruit. UHF tags are more expensive than low-frequency tags, and they con-sume more power.

Unique identifier (UID): A serial number that identifies a unique transponder. The U.S. Department of Defense has also developed an identification scheme called UID.

Wavelength: The distance between the same points on two consecutive waves. The distance can be measured from the peak or trough of one wave to the respective peak or trough of the next consecutive wave. Wavelength (λ) and frequency (f) are always related by the speed of light (c), such that $\lambda = c/f$.

WORM (write once read many): A type of RFID tag that can be written to only once, and subsequently can only be read. Usually associated with Class 1 EPC tags.

Index

• *N* •

• *O* •

• S •